José Fernandez

Colour profiles
Teodor Liviu Morosanu

POTEZ 63 Family

STRATUS

Published in Poland in 2008
by STRATUS s.c.
PO. Box 123,
27-600 Sandomierz 1, Poland
e-mail: office@mmpbooks.biz
for MMP,
36 Ver Road, Redbourn,
AL3 7PE, UK.
e-mail: rogerw@mmpbooks.biz

http://www.mmpbooks.biz

ISBN
978-83-89450-65-4

Editor in chief
Roger Wallsgrove

Editorial Team
Bartłomiej Belcarz
Artur Juszczak
James Kightly
Robert Pęczkowski

Colour Drawings
Teodor Liviu Morosanu

Scale plans
Marek Ryś

DTP
Artur Bukowski
Bartłomiej Belcarz

Translation
Patrick Laureau

Proofreading
Roger Wallsgrove
Matthew Willis

Printed by:
Drukarnia Diecezjalna,
ul. Żeromskiego 4,
27-600 Sandomierz
tel. +48 (15) 832 31 92;
fax +48 (15) 832 77 87
www.wds.pl marketing@wds.pl

PRINTED IN POLAND

Table of contents

Acknowledgements

Author and publisher would like to thanks following people:
J. Rajlich, T. L. Morosanu, F. Tulasne, P. Rouxel, J. Gaillard, C. Becze.

All photos from the author's collection, unless otherwise noted. Other photo credits: AM – Andrzej Morgała, BB – B. Belcarz, BK – B. Kimlička, EK – E. Knapová, FT – F. Tulasne, JA – J. Adam, JCS – J. C. Soumille, JR – J. Rajlich, JS – J. Szymankiewicz, MA – M. Andrei, PI&SM London, PR – P. Rouxel, Stratus – Wydawnictwo Stratus, TK – T. Kopański, WM – W. Matusiak.

Genesis and Development

This light twin-engine aircraft was developed and produced in multiple versions to equip a great number of units of the French *Armée de l'Air*, as well as some *Aeronavale* ones, in May 1940. In total, some 1,150 aircraft were built, and it was the most numerous combat aircraft to wear French colours.

The first wind tunnel model was a single fin aircraft.

The model of the first prototype with the flat horizontal tailplane.

A model of the Potez 63.0 with the new dihedral tail unit, to be installed when the prototype returned to the workshops after the accident on May 6th 1936.

Four views of the first prototype, the Potez 630 c/n 01.

The second prototype, the Potez 630 c/n 02.

The requirements published by the Air Ministry on October 31st, 1934, requesting a light defence multi-seat aircraft considered three kinds of mission:

- Fighter Unit management, in C3 configuration (three seats), carrying the commander of a single-engine fighter unit in order to direct them by radio
- Daylight escort fighter, in C2 configuration
- Night fighter, in CN3 configuration

These requirements included a 450 km/h (280 mph) maximum speed at 4,000 m (13,123 ft), this height to be reached in under 15 minutes. The cruise speed must reach 320 km/h (199 mph), with a four-hour endurance.

The armament must include at least two 20mm cannon as well as a machine-gun on a flexible mount for rear sector defence. The aircraft must not weigh over 3 tonnes (6,614 lb) in order to enjoy a level of manoeuvrability similar to that of a single-seat fighter, though this limit was later raised to 3.5 tonnes (7,716 lb).

Several companies responded to this tender. The main projects in competition were the Hanriot H.220 (later to be the N.C.600), Loire-Nieuport 20, Romano 110, Breguet 690 and Potez 63. The latter was proposed with Hispano-Suiza or Gnome-Rhône radial engines, resulting in the Potez 63.0 (or 630) and 63.1 (or 631). Potez produced a design by engineers Coroller and

The prototype of the Potez 630 during test flight.

The Potez 630 c/n 2.

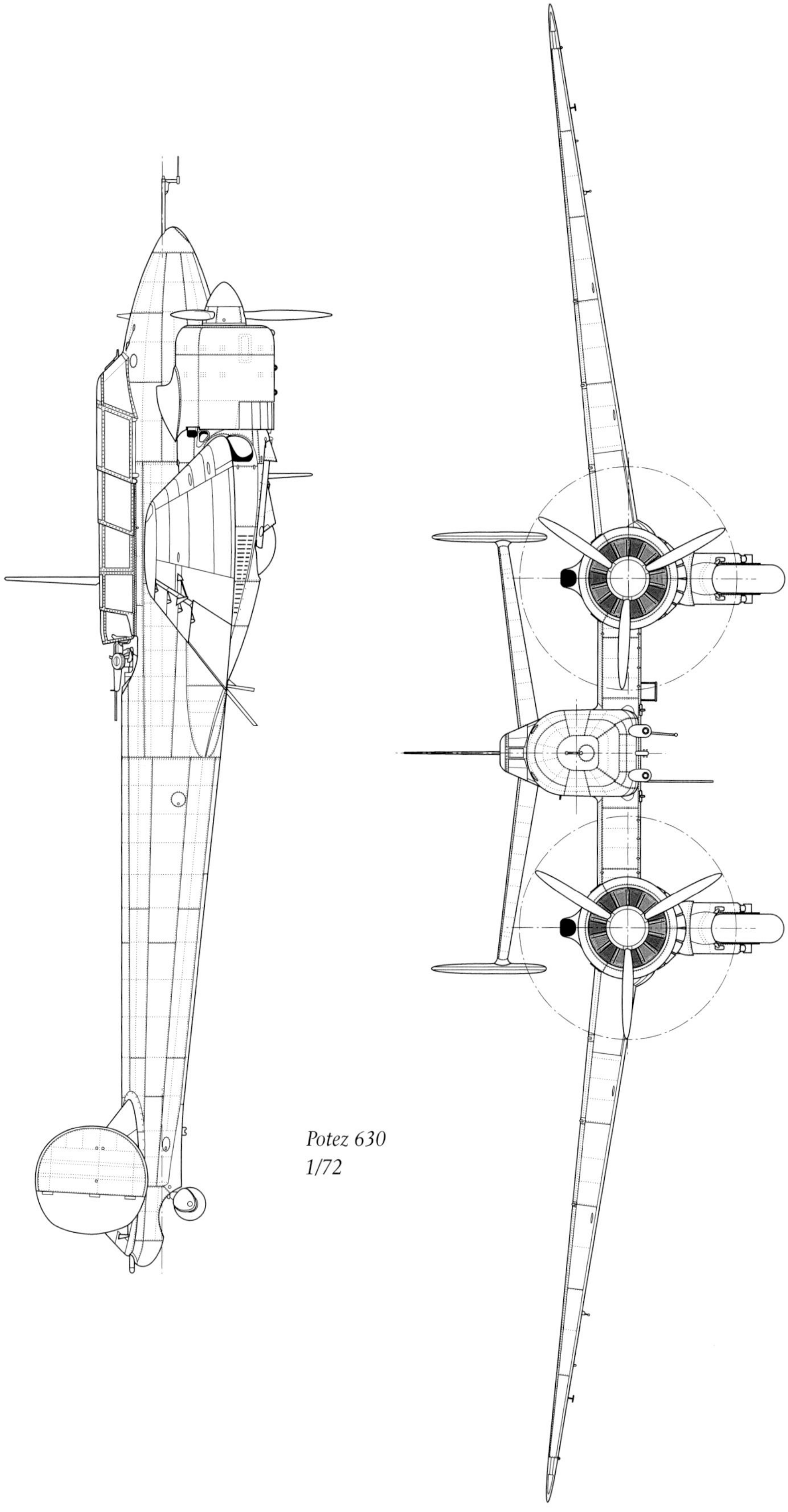

Potez 630
1/72

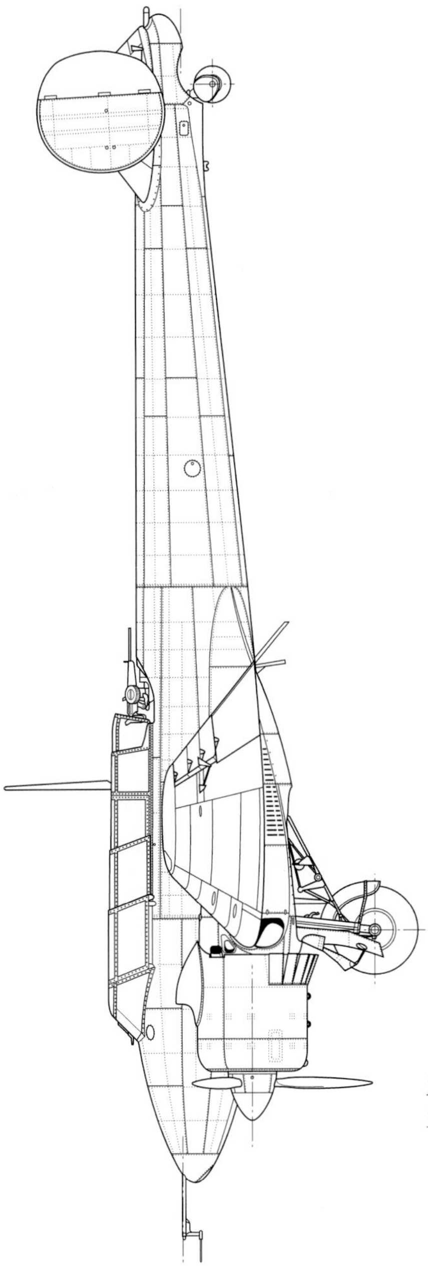

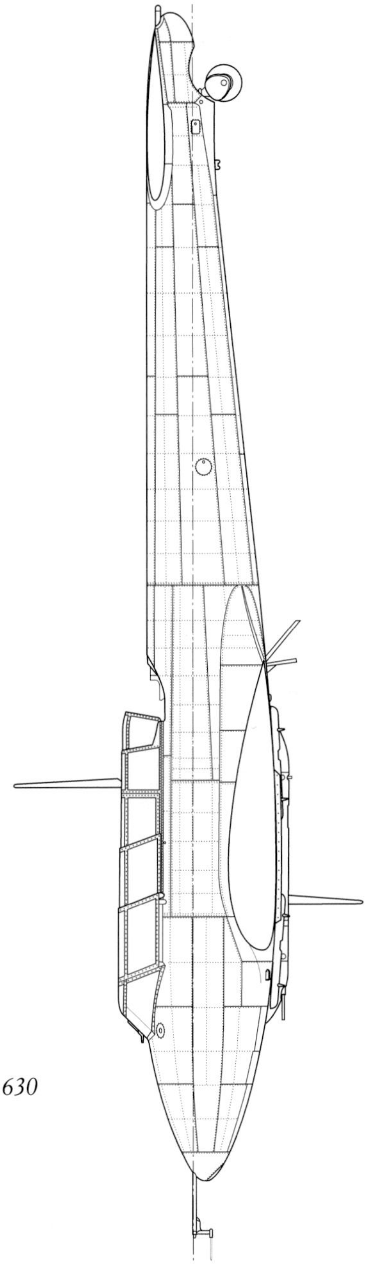

Potez 630
1/72

The first prototype of the Potez 631, c/n 01.

The Potez 631 c/n 01.

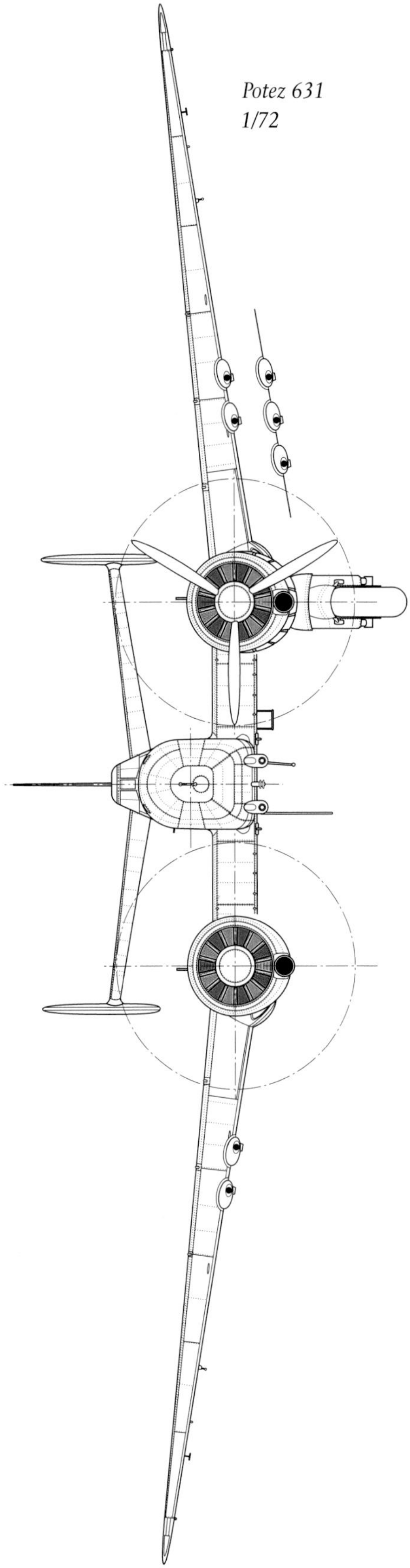

Potez 631
1/72

Delaruelle. The company began the construction of a first prototype, the Potez 630-01, in April 1935. It was then usual that the aircraft constructor paid for the project from its own funds, with the hope of being refunded in case of success.

The Potez 630-01 made its maiden flight on April 25th 1936, over the Méaulte plant's aerodrome in the hands of Frédéric Nicole, with 580hp Hispano-Suiza radial engines. On May 6th, when returning from a test flight, the aircraft experienced problems with its port engine, lost one propeller and was damaged on landing. Once repaired, it was sent to the CEMA (Centre d'Essais du Materiel Aerien) test centre at Villacoublay on August 3rd. These repairs were accompanied by a few modifications to the original airframe: the flat tailplane was replaced by a new model with dihedral, and the undercarriage legs were lengthened. The flight testing ended on November 20th, 1936. The engines were Hispano-Suiza 14AB 02/03 14 cylinder two-row radial units, delivering 640 hp at sea level and 725hp at altitude. During tests, the fully loaded aircraft reached 480 km/h (298 mph) at 5,000 m (16,400 ft) with an all-up weight of 3,850 kg (8,488 lb). Its climbing time to 4,000 m (13,123 ft) was 5min 56sec, its radius of action at 300 km/h (186 mph) cruising speed was 1,300 km (807 miles). The performance of the prototype thus matched the requirements published by the Ministry.

The aircraft was then delivered in late December to Villacoublay for the official test programme, led by military pilots.

Inside the hangar of the Meaulte plant, 1939.

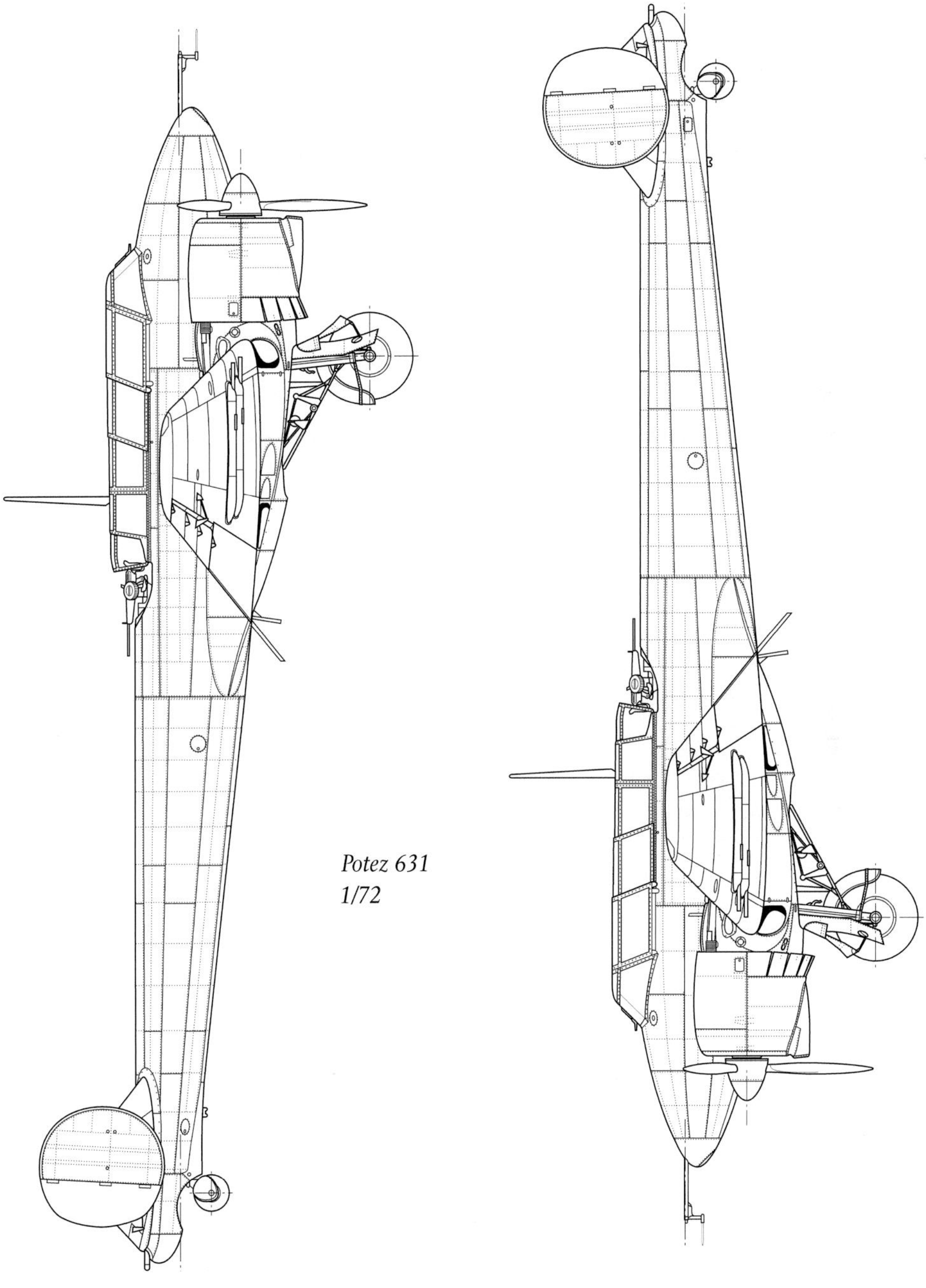

Potez 631
1/72

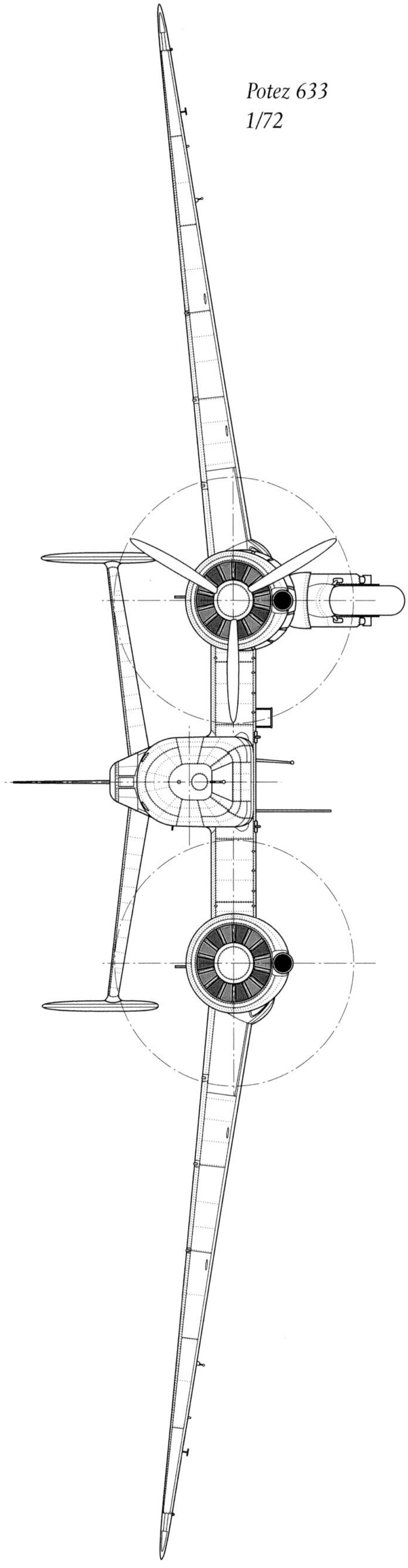

Potez 633
1/72

The second prototype was finished with Gnome-Rhône 14 Mars radial engines delivering 750hp at sea level and 660hp at altitude. The Gnome-Rhône units were less powerful than the Hispano-Suiza engines at altitude, but they were also much less bulky, with a low frontal area, and thus generated less drag. The 631 made its maiden flight at Méaulte in early March 1937, but was also damaged during landing because of a faulty undercarriage lock, and had to undergo some repairs. It was eventually delivered to CEMA for its flying tests, to begin in November 1937.

It was now SNCAN which managed the programme. The *Société Nationale de Constructions Aéronautiques du Nord* was created on December 23, 1936, in the wake of the nationalization and restructuring of the French aeronautical industry that followed the victory of the *Front Populaire* at the 1936 elections. During the first months of 1937, SNCAN would progressively swallow five plants and 5,496 employees, working at the Potez installations of Méaulte and Sartrouville as well as the plants of Les Mureaux, Caudebec and Le Havre.

In November, SNCAN received a letter of intention regarding ten aircraft earmarked for testing, all to be developed from the Potez 630 airframe: four more Potez 630s, three Potez 631s (including the prototype under evaluation), two Potez 633 B2s (a two-seat light bomber) and one Potez 637 A3 (a three-seat observation plane). Once the final contract was signed, one of the Potez 633 aircraft was replaced by a Potez 639 AB2 (ground attack two-seater).

The Potez 633 c/n 01, the prototype of the bomber version.

Potez 633
1/72

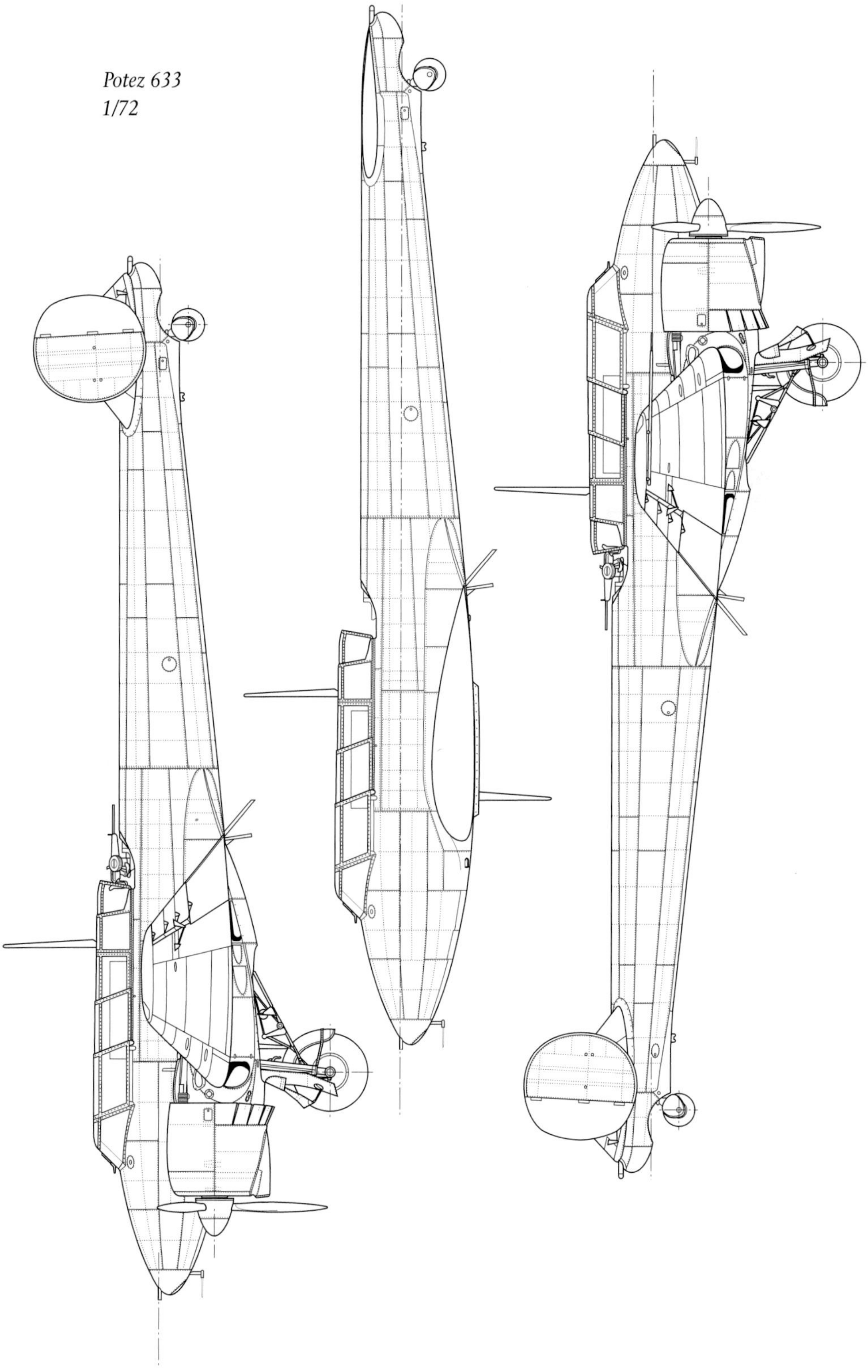

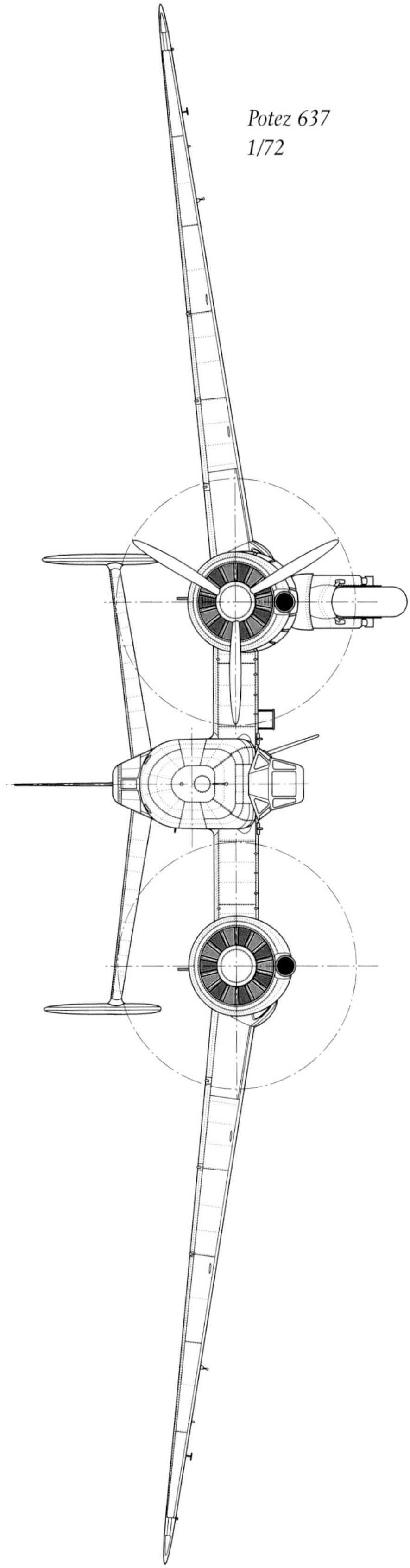

Potez 637
1/72

Production

The first production order was placed in June 1937 as a letter of intention, including ten two-seaters for training purposes (in order to convert crews to the new aircraft) and 30 Potez 631 three-seat fighters with the first aircraft to be delivered eight months later. However, the Gnome-Rhône engine production line was unable to match the airframe line, so a second order was issued for 80 Hispano-Suiza equipped aircraft (Potez 630). The engines would be the 14 AB 02/03 on 48 of them, and the 14 AB 10/11 on the remaining 32 aircraft. In December 1937, a new order was issued for 50 Potez 633 B2 bombers, bringing the total to some 180 aircraft, including the ten prototypes.

During 1938 the new aircraft raised some interest abroad. The Czech company Avia acquired a licence to manufacture the Potez 636 three seat fighter variant. However, this was nullified by the German annexation of Czechoslovakia. Meanwhile China, Yugoslavia, Romania, Greece and Switzerland all issued orders for distinct versions of the Potez 63.

The production of the Potez 63 started in May 1937. The fuselages and tail units were built in Caudebe c, the wings in Le Havre, and the final assembly was performed first in Méaulte, then in Les Mureaux after the completion of the last batch of MB-210 bombers.

The first production aircraft was out of the assembly line and made its first flight at Méaulte in February, 1938.

At Meaulte plant, the last step of work, painting and varnishing.

Potez 637
1/72

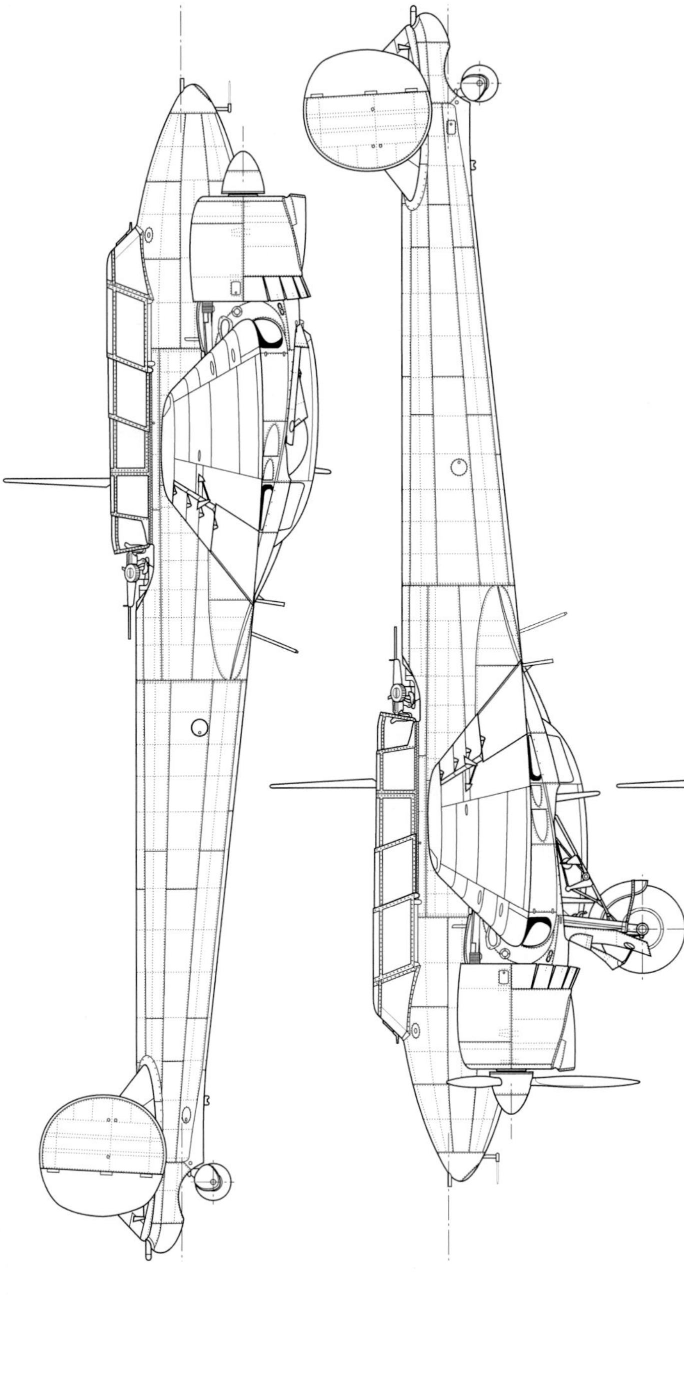

Potez 63s in a hangar at Meaulte, September 2nd 1939. Nine of them are Potez 630s (c/n 67, 13, 14, 15, 16, 17, 18, 19) of which seven will be knocked down and crated for shipment to China.

Inside the hangar of the Meaulte plant, 1939.

The *Armée de l'Air* received Potez 630 c/n 8, the first aircraft to enter service, on May 23, 1938, and a further three were delivered later that month. C/n 8 underwent various tests and reached 448 km/h (278 mph) at 4,000 m (13,123 ft), which it reached in 7min 1sec. On August 2, 1938, the first Potez 631 (c/n 5) was delivered and climbed to 4,000 m in 5min 5sec, with a maximum speed of 437 km/h (272 mph).

The assembly line at the Meaulte plant.

The 5th Plan for re-equipment of the *Armée de l'Air* was initiated on March 15th 1938. It considered the adoption of one of the twin-engine projects born from the 1936 requirements, but with far better performance than the Potez 63. Unfortunately, none were sufficiently developed to be put into production rapidly.

As a consequence, the Potez 631 and 633 were to be included in the new production programme. New orders arrived, for 207 Potez 631s and 449 Potez 633s (the latter figure changed to 453 some ten days later). During May the Headquarters decided that the new light bomber must be a three-seater, and the Potez 633 order was thus switched to Potez 631. In the meantime it was also decided to withdraw the Potez 630 from service, because of the problems with the Hispano-Suiza engines; they would be converted into training aircraft for crew conversion.

The Potez 637 A3 C/n 1 prototype was under construction. It was a reconnaissance version fitted with a belly gondola, and flew for the first time in October 1938. But the prototype had not been fully assembled when the first order

This Potez 630 (French registration F-AREY) is one of the aircraft sold to Yugoslavia.

was issued for 60 Potez 637s, to replace as soon as possible the Potez 452 still in service in four reconnaissance groups.

The *Armée de l'Air* was faced with the replacement of the aircraft equipping the *Groupes Aériens d'Observation*, two-seat single-engine aircraft such as the Breguet 27, Potez 39 and Mureaux 115 and 117, which were all completely obsolete. Precise requirements were studied regarding a three-seat observation aircraft destined for army cooperation.

As a consequence, SNCAN launched the development of a new observation aircraft derived from the Potez 63, the Potez 63-11. Three prototypes were ordered, and before the end of 1938 three different orders would total 465 Potez 63-11s.

In the meanwhile, the delivery of the Potez 630 and 631s was continually postponed because of delays in the production of engines, propellers and guns. The first batch of Potez 630s was armed with four machine-guns instead of the two 20mm cannon. The first estimates considered some 201 Potez 63s in service on January 1st, 1939, but on that date the *Armée de l'Air* had only received 45 Potez 630s and 27 Potez 631s, while 74 complete airframes were on the lines waiting for their engines. Among the aircraft delivered, sixty were provisionally fitted with Levasseur two-blade wooden propellers, rather than the specified 3-blade metal units, so that the air force could take them on charge. This allowed aircraft to be put in service without having to wait for propellers, and in spite of all the problems, some fifteen Potez 63s took part in a parade over Villacoublay in July 1938. In September, 17 Potez 630s replaced the MB 200 bombers used as command aircraft by the fighter wings at Reims, Dijon, Etampes and Chartres.

By April 1st, 1939, the *Armée de l'Air* had received 77 Potez 630s and 88 Potez 631s, of which 67 and 20 entered service in fighter units respectively. Each single-engine fighter unit received six Potez 630s, as did the night fighter units equipped with Mureaux 113 CN2s (GCN III/1 and II/4, later to be GCN I/13 and II/13). Even so, the new aircraft were rapidly replaced by Potez 631s. On the other hand, GC II/8 (Morane-Saulnier MS 225 and Dewoitine D 510) was entirely re-equipped with 18 Potez 631s.

At the end of the month, a unit of four Potez 631s was set up at Marignane within GAO 550 (renamed as GAM 550, for *Groupe Aérien Mixte*). In August, another unit of the same type was established in Djibouti.

In May 1939 the first Potez 637s reached their units. GR I/33 and II/33 and GR I/52 and II/52 based in Nancy were also re-equipped in the same way. The war was close, and the situation of the Potez 63 in France was as follows:

Potez 630: 85 aircraft delivered, 65 in service in combat units

- 8 aircraft with the 1st Fighter Wing in Etampes
- 12 aircraft with the 2nd Fighter Wing in Chartres
- 9 aircraft with the 3rd Fighter Wing in Dijon
- 1 aircraft with the 4th Fighter Wing in Reims
- 8 aircraft with the 5th Fighter Wing in Reims
- 8 aircraft with the 6th Fighter Wing in Chartres
- 10 in Tunisia, including 3 with GAC in the capital

Potez 631: 206 delivered, 117 in service in combat units:

- 18 aircraft with GCN I/13 in Etampes
- 21 aircraft with GCN II/13 in Reims
- 4 aircraft with the 1st Fighter Wing
- 7 aircraft with the 2nd Fighter Wing in Chartres
- 6 aircraft with the 3rd Fighter Wing in Dijon
- 8 aircraft with the 4th Fighter Wing in Reims
- 6 with the 5th Fighter Wing in Reims
- 7 with the 6th Fighter Wing in Chartres
- 6 with the 7th Fighter Wing in Dijon
- 26 aircraft with the 8th Fighter Wing
- 4 aircraft with GAM 550 in Ajaccio (Corsica)
- 4 with the Colonial Squadron in Djibouti

Potez 633: 22 aircraft delivered (3 from the Chinese order and 19 from the Romanian order) including 3 in service with GB I/54 on September 1st, 1939.

Potez 630 of GC I/5 (Fighter Group). It has just been delivered to the unit base of Reims.

First of this line-up, Potez 630 c/n 9.

Ground engine run of a Potez 630.

Potez 630 c/n 2 is overflown by another one while being refuelled.

A Potez 630 crew climbing aboard. This camouflage was painted on almost all Potez 630s of the first batch delivered to the Armée de l'Air.

Another Potez 630 having its engines ground run. Note the armament under the fuselage: a machine gun and a cannon.

A Potez 630 with the upper surfaces in French khaki colour.

This Potez 630 shows underside details.

A Potez 630 being pushed, seen from under a LeO 20 bomber.

Potez 630 c/n 44 of GC III/1 in three-tone camouflage with 30 cm roundels on wing uppersurfaces.

Potez 630 from 3rd. Escadrille GC II/1, later this airplane was transfered to Chartres, where it was used for training by Czech pilots. In the photo, por. *Karel Vykoukal and* por. *Rudolf Roháček.*

BK via JR

Potez 630 c/n 13 (C-525) of GC I/2. It is a "commanding aircraft for the fighters"

Potez 631 in flight.

Potez 637: 61 aircraft delivered, 53 in combat units:
- 26 aircraft with the 33rd Reconnaissance Wing in Nancy
- 27 aircraft with the 52nd Reconnaissance Wing in Nancy

Potez 63-11: 5 aircraft delivered, none in combat units

The French air arms had required a crash rearmament programme, and the aeronautical industry had faced huge difficulties in producing the modern aircraft needed to replace the obsolete planes still in service on the eve of war. In spite of these difficulties, and limited resources, the Potez 63 looked like a winner. Its construction needed only 7,500 hours of work, the same amount as the D 520 single-seat fighter, or half that needed to make one Morane-Saulnier MS 406. It was therefore able to be produced in large numbers at a good rate.

Por. *Vladimir Zaoral in front of a Potez 630, Chartres.*

JR

The Phoney War

In Fighter service

On the eve of the war *Escadrille de Guet* I/16 (Watch Squadron) at Clermont-les-Fermes (which would become the *Escadrille de Chasse Multiplace*, or ECM I/16 on January 29th, 1940) had two Potez 630s and ten Potez 631s, drawn from the assets of GC II/2, GC I/6, GC II/4, GC II/5, GC II/6 and GC II/7.

On October 16th, 1939, the 5th Night Fighting Squadron was set up at Lyon-Bron with twelve Potez 631s. It became fully operational on January 1st, 1940, as ECN 5/13. GCN I/13 and II/13 were divided into four independent squadrons, called 1/13, 2/13, 3/13 and 4/13.

The *Aeronavale* received its first batch of Potez 631s in January 1940. These aircraft were transferred from the *Armée de l'Air* and were destined for *Escadrille de Chasse* AC1, which had disembarked from the aircraft carrier *Béarn* with AC2 a few months before. Both were equipped with obsolete Dewoitine D.373 and 376 single-seat fighters, and AC2 would also receive Potez aircraft some time later. In March 1940, while the crews converted onto the new twin-engine fighter, the 1st *Flottille de Chasse* F1C was created, to be based in Calais, and regrouping AC1 and AC2. The *Flottille* received 17 Potez 631s from GC II/8 (which had received Bloch MB 152 single-seat fighters in March and April). By mid-May, the unit was fully equipped with 26 Potez 631s.

In February 1940, after all fighter units' airborne command sections were disbanded, the last Potez 630s were withdrawn from the *Armée de l'Air*'s operational units. They were subsequently transformed into dual control training aircraft. The night fighting units were reinforced with their complement increased from 12 to 18 Potez 631s per squadron.

In February 1940, it was also decided to reinforce the Potez 631's armament, which had proved insufficient - most of them had only one 20mm cannon and one 7.5mm machine-gun, instead of the planned two 20mm cannon. Taking advantage of routine maintenance cycles, the aircraft would finally receive their two 20mm cannon, plus four extra machine-guns under the wings. It was thought

This Potez 637 (c/n 38) crashed after a reconnaissance flight over the front line. Autumn 1939.

Two photos of Potez 637 damaged during landing.

that this measure was would give the aircraft a genuine air-to-ground capability. However, by the time of the German invasion on May 10th 1940, only two aircraft had been modified, both being in service with ECM I/16.

Light Bomber

Delivery of the Potez 633s under production for foreign orders was embargoed, and the aircraft were delivered to *Armée de l'Air* units. GBA (*Groupes de Bombardement d'Assaut*) I/54, II/54, I/51 and II/51, though waiting to be equipped with Breguet 691s, received these 633s in order to train the crews, until the arrival of the Breguet 691 and 693. In December 1940 *Groupements d'Assaut* 18 and 19 of the 6th *Brigade de Bombardement d'Assaut* based in Vinon were also equipped with Potez 633s, as GBA II/35 in March 1940. The crews of these units flew some 40 to 60 hours on Potez 633s before switching to their Breguets.

During the Winter War (November 1939 to March 1940) between Finland and the USSR, an expeditionary unit was planned (called *Groupe Finlande*) in order to support the Finns. It was set up in Vinon with a dozen Potez 633s and voluntary crews from the 54th Wing. The unit flew to Tangmere in England on March 11th 1940, where it joined British units formed for the same purpose, but the adventure went no further, as an armistice put an end to the hostilities between the Finns and the Soviets until June 1941.

Potez 630 c/n 82 from Groupement Provisoire d'Instruction de la Chasse *on Oran-La Senia airfield. On wings:* por. *H. Hrabáček. Standing from left:* por. *B. Tobyška,* por. *F. Sýkora,* čet. *J. Adam,* čet. *J. Nowák,* por. *S. Janouch.* ***EK via JR***

Potez 631 c/n 140 of ECM I/16.

Potez 637 c/n 45. The first war mission for GR I/52 was carried out by Cne. *Bertaux from Chaumont-Semoutiers on September 4th, 1939.*

Reconnaissance

When war was declared, most of the *Groupes de Reconnaissance* were still flying Potez 540/542s and Bloch MB 131s, or even Potez 25s, in the case of GR II/39 in Syria. The Potez 637 was the only modern reconnaissance aircraft available, and constituted the backbone of four Reconnaissance Groups (which retained some obsolete Potez 540/542s for training purposes). These were:

- GR I/33 based in Saint-Dizier
- GR II/33 based in Saconin
- GR I/52 based in Chaumont
- GR II/52 based in Herbeviller.

Potez 637s used for reconnaissance missions over the front lines rapidly had their baptism of fire. On September 8th, 1939, a Potez 637 (c/n 52) from II/52 was intercepted by *Lt.* Gutbrot from II/JG 52, and was seriously damaged but reached its base flying on one engine. On September 20th, an aircraft from I/33 was shot down by *Flak* while flying over enemy positions, and its crew were captured. Between September 24th and 30th, five Potez 637s were lost, together with five crews killed and another two wounded. At the end of the war, seven other aircraft would also be counted as lost.

The Potez 63-11 was beginning to be delivered to some units, and is considered below. These aircraft also reinforced the Potez 637 units, the latter being progressively replaced and transformed into dual control trainers intended for the training of Potez 63-11 crews. The Potez 637 may have been somewhat faster and more manoeuvrable than the 63-11, but the latter was considered to be better adapted and more efficient to the reconnaissance missions it had been designed for.

Crash-landed Potez 631 c/n 172.

The French Campaign

On May 10th, 1940, Germany invaded France. The situation of the Potez 63 units was as follows:

Fighters

As the Potez 630 had been withdrawn from service, the Potez 631 was the only version in operational units. In the north, at Calais-Marck, was based the *Aeronavale*'s *Flottille* F1C including AC1 and AC2. At Wez-Thuisy was ECMJ 1/16, part of the *Groupement de Chasse* n°23.

Night defence of the Paris area was entrusted to the *Groupement de Chasse de Nuit* including four units: at Le Plessis-Belleville ECN 1/13, at Meaux-Ebly 2/13, at Melun-Villaroche 3/13 and at Betz-Bouillancy 4/13.

Bombers

There were only about twenty Potez 633s left in combat units. All were included in the squadrons of the 6th *Groupement d'Assaut*, undergoing conversion to the Breguet 693, and distributed among GBA I/51, II/51, I/54 and II/54.

A Potez 631 during overhaul in summer/autumn 1940, after the armistice. In front of two Bloch 210s and three Curtiss H-75s. Writing on the wall reads: smoking prohibited.

Reconnaissance

The Potez 637s still in service flew together with the Potez 63-11 and Bloch 174. These new aircraft were re-equipping the units, whose crews were being converted.

GR II/33 was based at Athies-sous-Laon, to carry out the strategic reconnaissance needs of the 1st Air Division. GR II/52 was at Couvron, to fulfil observation and reconnaissance missions for the 9th Army. GR I/52 was at St-Dizier, for the strategic reconnaissance needs of the 3rd Air Division. GR I/33 was at Dôle, doing the same for the 6th Air Division. The other reconnaissance and observation units had already received their Potez 63-11s.

Potez 637 c/n 47, 1. Escadrille GR I/52, damaged September 6th 1939.

Potez 631 c/n 191 sports the small-sized roundels on its wings.

The wreckage of one of the Potez 631s shot down during the French campaign.

Sgt. *Pierre Rouxel, of Groupe de Chasse III/2, dressed in a high altitude flying suit, in front of his Potez 631.*

PR

Combat

Fighters

The Potez 631 was quite successful against the *Luftwaffe* in the hands of the *Flottille* F1C pilots who obtain a dozen victories between May 10th and May 21st 1940 for the loss of eight of their aircraft. They operated over the North Sea and Nord-Pas-de-Calais region for eleven intensive days. Following this period, the *Flottille* was sent to the rear to convert to the Dewoitine D 520 and Bloch MB 151 single-seat fighters, but the armistice was be signed before it could return to combat. Lt.V. de Scitivaux, one of the F1C pilots wounded on May 10th, eventually escaped to Great Britain and resumed service with the Free French Air Force (FAFL).

The ECMJ 1/16 Potez 631s had a harder time against the *Luftwaffe* during their daylight fighter missions. On the 13th, a Dornier Do 17 was damaged by *Cne* Escudier, though he could not shoot it down as the speed difference did not allow him to keep the Dornier within range. Some time later, when four aircraft simultaneously attacked a formation of He 111s, two of the latter were shot down, but during the day *Cne.* Escudier was forced to make an emergency landing after being hit by the fire of several RAF Hurricanes! The following day, another Potez 631 was engaged by Hurricanes, and this with even worse results, as *Sgt.* Simon was killed and *Sgt.* Sauvage had to bail out.

The units responsible for the defence of Paris were thrown into ground attack operations. Around Fourmies, 24 Potez 631s attacked German columns at the same time on May 17th, and two aircraft were seen to fall in flames after being hit by *Flak* while four other aircraft failed to return. *Cne.* Petit de Myrbeck, 2/13's CO, was missing, and a crew member of 1/13 was killed.

The first victory for the night fighter units was obtained on May 18th against a He 111 by an aircraft from ECN 1/13, flown by *Sgt/Chef.* Post, *Cne.* Treillard and *Adj.* Guichard.

Potez 631 c/n 164 of ECN 1/13 based at Nîmes-Courbessac in August 1940, seen here during a training flight. Above the unit's emblem (a white trident inside a blue pennant) it carries the name Sgt/Chef. *Collinot, in memory of an airman killed in action during the French campaign.*

***Left and below:** German soldiers inspecting Potez 631 c/n 146 of ECN 4/13, shot down above Chavigny on May 27th, 1940, by Messerschmitt Bf 109s.*

As demonstrated by the cases of *Cne.* Escudier and *Sgt.* Simon, confusion was common on the part of Allied aviators as well as anti-aircraft artillery. The Potez 63 was often mistaken for the Messerschmitt Bf 110, as the two fighters' silhouettes were very similar. On May 23rd a Potez 631 of ECN 3/13 was shot down by Bloch MB 152s from GC I/1. In order to avoid this kind of confusion, a special marking was adopted by all Potez 63s - a wide white line painted on each fuselage flank, while the fuselage roundels gained a white border.

Altogether, the Potez 631 fighter units obtained 17 confirmed air-to-air victories during the French Campaign which, added to the 12 victories of the *Aeronavale*, amounted to 29 successes for the Potez 631. As for the ECNs, the total was only four confirmed victories plus eight probables, against ten aircraft lost, eleven crews killed and another eleven wounded.

ECMJ 1/16 obtained three confirmed victories plus one probable, against seven aircraft lost, two crews killed, six wounded and two prisoners.

Bombers

When the Germans launched their offensive, there were still a few Potez 633s in operational service alongside the more common Breguet 693s. GBA II/51 was transferred to Etampes, where it joined on May 19 with GBA I/51, in order to reinforce *Groupement* 18. The unit was involved in continuous attacks against the German columns between May 20th and 30th, and lost six aircraft. C/n34 was shot down by *Flak* and its crew captured, c/n 36 was hit by enemy fire and destroyed on landing at Etampes airfield, and c/n 74 crashed on May 30th, south of Amiens. On May 22nd there were still ten operational Potez 633s (c/n 12, 17, 18, 21, 23, 24, 27, 31, 41 and 50) in service with GBA II/51 and II/35.

Potez 637 c/n 22 of GR I/33. Aircraft was painted in special markings on May 24th. These markings were introduced to distinguish the Potez from the Bf 110.

A Potez 631 of GC III/2 on its airfield in autumn 1939. ***PR***

A Potez 630 of GC III/2, with Sgt.Rouxel at the controls.

PR

Potez 631 c/n 48, GC II/8, after a forced landing during the winter of 1939-1940.

JCS coll.

Preflight checks for Potez 631 c/n 204 of GC III/2, before a reconnaissance mission.

PR

Unknown Potez 637 with crew before combat mission.

Reconnaissance

The reconnaissance Potez 637s also paid a high price. The only unit exclusively equipped with the 637 was GR II/52, which lost all its aircraft in only sixteen days of operations. At the end of May it was completely re-equipped with Potez 63-11s, and on June 17th it flew to Bordeaux to convert to the new Bloch 174, the aircraft it was expected to operate from the beginning, but whose deliveries had been continuously postponed.

On May 10th GR II/33 included some seven Potez 637s, six Potez 63-11s and four Bloch 174s, and received two Potez 631s as reinforcement from ECMJ 1/16 on May 18th. At the end of June, the group only had 5 Bloch

Potez 637 c/n 25 of GR I/33, shot down by German Flak on September 20th 1940.

Another view of Potez 631s being overhauled during summer/autumn 1940, after the armistice. Bloch 210s, Br 693s and a Leo 451 in background.

174s at Alger-Maison Blanche airfield, and all the Potez 63s had been lost in combat or abandoned on various airfields during the unit's retreat.

All the Potez 63s had hugely suffered during their low altitude attack missions against the German mechanized columns and their efficient *Flak* defences. The lack of air cover from Allied fighters also turned them into easy targets for an omnipresent *Luftwaffe*. Many aircraft were also destroyed on the ground by the numerous enemy raids. In addition to all these factors, a lack of spares made repairs very problematic. The number of available reserve aircraft was small and insufficient to cover attrition. In May 1940, there were only 9 Potez 631s in reserve, and in June only six.

After the Armistice

In the so-called French Free Zone were the following aircraft:

- 32 Potez 630s
- 112 Potez 631s
- 18 Potez 633s
- 12 Potez 637s

In North Africa, there were 53 Potez 63s of all kinds. When the Vichy regime organized its own aviation units, the surviving Potez 63s were integrated into the new units, and on November 1st, 1941, they had 22 Potez 630s, 82 Potez 631s and 6 Potez 637s. By October 1942, there were only 6 Potez 630s and 64 Potez 631s.

A Potez 631 of Bretigny flight test center with an AAC-1 Toucan (French Ju 52) and a Nord light plane after WWII.

A pilot stands on a Potez 631's wing root in 1942. Engine cowl wears the infamous Vichy yellow-red stripes.

Potez 631 c/n 52, Vichy Air Force, 1941.

Night Fighter Squadron 1/13

during the 1939-1940 Campaign

This eye-witness report was written by **Sgt/Chef.** ***Post, one of the unit's pilots, during the autumn of 1940.***

Before the war, *Escadrille* 1/13 was based at Etampes. With 2/13, it constituted a night fighter group assigned to the defence of the Paris area. This group was commanded by *Cdt.* Geoffroy, the squadron by *Cne.* Treillard. There were nine crews, flying the Potez 631, a three-seat twin engine aircraft with an offensive frontal armament and a defensive gun on a flexible mount.

During the six months preceding the war, and thanks to the efforts of *Cdt.* Geoffroy, a demanding and energetic leader, as well as *Cne* Treillard, a vigorous leader, the crews reached a level of comprehensive training on their new aircraft. The technical staff, which had been considerably reinforced with the arrival of young mechanics fresh from school, was taking great care of the new planes. When the time came for combat, all the components of the squadron were ready.

On August 21st, 1939, considering the worsening diplomatic situation, the squadron was called back to Etampes from the night fighting exercises it was performing in Reims. The A echelon was mobilized. On Sunday, August 27th, it received the order to disperse to the aerodrome of Meaux-Villenoy, which would be its base of operations. The ground and air components were on site during the day, and the crews immediately flew the first reconnaissance flights over the area while the mechanics were organizing their facilities.

The crews were distributed as follows:

Aircraft c/n	Pilot	Gunner	Mechanic
117 (1)	*Cap.* Treillard	*Sgt*/Chef. Ségni	*Sgt*/Chef. Gillet
113 (2)	*Lt.* Pelleray	*Adj.* Saint-Aubert	*Sgt.* Besson
96 (3)	*Lt.* Audebaud	*Adj.* Gobert	*C/C.* Gaumin
42 (7)	*Lt.* Rousseau	*Adj.* Lièvre	*Sgt.* Montagnier
20 (4)	*S/Lt.* Boursain	*C/C.* Jobard	*Adj.* Jan
109 (5)	*Adj/Chef.* Laurent	*Sgt*/Chef. Lenoir	*Adj.* Godard
104 (6)	*Adj.* Hénaut	*Sgt.* Sannier	*Sgt.* Neusius
73 (8)	*Sgt.* Biger	*Sgt*/Chef. Boutillon	*Sgt*/Chef. Frey
32 (9)	*Sgt.* Post	*Adj/Chef.* Birembaut	*C/C.* Rozé

On August 28th, the role of the squadron was explained to the crews, who received all the information relevant to their task. The unit was included in the *Zone d'Opérations Aériennes du Nord*, and constituted one of the main components of Paris air defence. The commanding officer of the whole system of night fighter squadrons and searchlights was *Lt/Col* Dordilly, the group's former leader. The *Groupement Mixte de Nuit* n°2, under the command of *Cdt.* Geoffroy, was made of two squadrons, 1/13 and 2/13, and searchlight batteries. The sector it had in charge went from north to south with Meaux as an axis. It was linked with other systems through AA covered areas, and the complete defensive arrangement formed half a circle, east of Paris.

The air defences had to work together with the information centre, which transmits the alerts from the watch service, and with the help of the searchlight batteries, whose task was catching intruders within their light beams. Any air action was thus subordinated to these components. But during exercises, it rapidly became clear that delays in data transmission might make the crews' flights completely in vain. In fact, as the future would prove, the night fighters never had an opportunity to intercept the enemy in the air. In spite of this painful conviction, the unit showed an absolute trust and devotion to its duty. The first weeks of the war were difficult. The importance of the task meant that the squadron maintained a permanent alert system, and the crews had no relief, working day and night. Moreover, there were practically no facilities - the only shelter for the alert crews and mechanics was a field tent, the only beds available were pneumatic mattresses. All during the autumn and winter, under temperatures sometimes reaching -20°C, they waited every night for the take-off order, under these conditions.

The first mission took place on September 6th. *Cne* Treillard and *Sgt/Chef.* Séghi took off on alert at 01:50 on a totally blacked-out misty night, to cover the sector for forty minutes, until the end of the alert.

On September 11th, the unit received a visit from *Gen.* d'Harcourt, Inspector of the Fighter Command, who explained the situation of air operations on the French and Polish fronts, and provided them with the latest information regarding the tactics and armament used by the enemy.

The limited activity displayed by the enemy during the months preceding the offensive of May 1940 did not diminish the squadron's dedication. While the crews fulfilled their defensive task over Paris at night, they also performed a great number of training flights, night and day. These were mainly exercises in common with the AA and searchlight systems, which were notoriously deficient and the cause of serious concern.

Soon, the commander decided to train the crews for daylight fighter missions in order to be able to use them, should the need occur. The squadron continued with its night task, while starting instruction that included all the combat manoeuvres and situations, with an important role of aerobatics. Here, the Potez would reveal its manoeuvrability, a remarkable quality for a twin engine aircraft.

In October 1939, the already weak flying personnel was further depleted by the loss of *Adj/Chefs* Laurent and Birembaut, who were assigned to *Escadrille* 5/13. There were five officers and only one NCO left in the squadron, for nine aircraft available. At all levels, the CO increased his efforts to get new pilots.

The armament of the aircraft was notoriously insufficient, and constituted another of *Cdt.* Geoffroy's protests. This armament, one cannon and one machine-gun, was nevertheless not augmented before June 1940.

Between October and December, 1940, the squadron periodically detached two crews to Melun-Villaroche airfield, at disposal of the *Groupement Mixte de Nuit* n°3. They had to cover the southern sector in case of an alert, and simultaneously perform exercises with AA and searchlights. Their mission ended when 2/13 took possession of the base.

During the winter, enemy air intrusions were slowing down. Nevertheless, the crews kept training and flew in any weather. The pilots became proficient in combat and manoeuvring, and piled up night flying hours. Around Paris, new high power Sperry searchlights were tested, and the crews who witnessed their efficiency recovered some confidence in the possibilities of night fighting.

This was the spirit which prevailed while the squadron was waiting for a season that should, undoubtedly, bring up some memorable events. Of course, everybody was conscious of the limited means, but courage and resolution were also weapons, that all of us were wearing. At that time, the squadron

During spring 1940, S/Lt. *Boursain gets ready for a mission aboard Potez 631 C3 c/n 110 of 1/13.*

experienced death for the first time in years. On March 4th, 1940, after a night flight, a Potez 540 used as a target for night fighter exercises crashed on landing, killing two and with three wounded crew.

New crews came and reinforced the squadron. Two radio operators, *Cpl/Chefs* Dobsik and Pelletier, were thus integrated. Then on March 8th, *S/Lts* Merle and Giger, formerly NCOs in the group, were assigned to the unit and immediately resumed their training. The arrival of such two excellent pilots was highly appreciated. On the other hand, *Lt.* Rousseau, who had no night flying experience, was assigned to an instruction centre, and his leaving did not diminish the unit's potential.

On March 9th, two pilots from *Groupe Mixte de Nuit* n°1 came and reinforced the squadron with one extra aircraft. But this plane was destroyed in a bad landing, and the two pilots went back to their base on the day following their arrival.

At the end of March, pilots and gunners performed air-to-air firing against towed targets at Mourmelon. The results revealed sharp shooters among all the units, but 1/13 obtained the best results.

On March 23rd, two brand new Potez 631s were added to the squadron's fleet, now reaching eleven planes. In the following weeks, the crews trained in night flying navigation, and flew in formation through very dense cloud layers before gathering over them. These exercises complete the daylight fighting training and gave the pilots new expertise, enabling them to enter combat in the best possible conditions.

In early April, new promotions occurred among the NCOs. Then *Adj.* Voirin, *Adj/Chef* Blanc and *Sgts* Soulié and Cailleux, who come from a heavy fighter unit in Calais, were assigned to the squadron as gunners. On April 15th, *Lt.* Pelleray was promoted to *Cne.*, and on the same day, *Cdt.* de la Perelle replaced *Cdt.* Geoffroy at the head of *Groupement Mixte de Nuit* n°2. As the new commander left his Hanriot 182 liaison aircraft at *Cne.* Treillard's disposal, the latter took a brilliant decision - if the squadron cannot get new pilots, the squadron will instruct his own pupils! And thus *Sgt/Chef.* Post was promoted as an instructor and began flying lessons for gunners and radio-operators.

Over nine months had passed since September 2nd, 1939, but the lack of action had not diminished the courage of the crews. On May 10, the general alert rang at 04:30. The unit was told about the invasion of Holland, Belgium and Luxemburg by the German armies. All were ready to fulfil their duty. They will not be spared...

Until then, the crews had accumulated thirty-three wartime night fighting missions on alert. But the enemy did not fly much at night, and considering the serious insufficiency of French aviation, the high command will not let them rest. As from then, they will take part in the great air and ground battle, and fly their "bat" wherever combat is.

The Potez 631 was too slow for a fighter, badly armed for assault, badly equipped for reconnaissance. Does that matter? It was badly needed. It will thus perform all these missions. Between May 10th and 16th, two small flights (all that was available!) will cover all the sectors. No combats were registered against the enemy, who gathered his forces on other fronts.

On May 13th, three crews from Squadron 4/13 were temporarily added to 1/13, and took their share of the cover missions. Soon the advance of the *Panzerdivisionen*, breaking through the fortified lines in several places, created such a great danger that all means had to be used to stop them. The assault aviation units had been virtually annihilated during the previous days, and cannot fly anymore. The night fighter units will replace them!

On May 17th, all the units were ordered to attack the enemy bulge generated north of Hirson. The target of *Escadrille* 1/13 was the road from Trélon (France) to Chimay (Belgium), and on which the enemy elements were travelling.

Two flights took-off from Meaux at 08:00, crewed by *Cne* Treillard*Adj.* Séghi, *Sgt/Chef.* Post, *S/Lt.* Merle, *Adj.* Boutillon, *Sgt.* Sannier, *Cne* Pelleray, *Adj.* St Aubert, *S/Lt.* Giger, *S/Lt.* Bourdain and *Adj/Chef* Gobert, *Sgt.* Jobard. They were followed by a third flight with three crews from *Escadrille* 4/13. Soon St Quentin was behind them, when the French AA opened fire, without hitting them. Then the target was identified, and each flight successively attacked the columns, firing with all their guns. The enemy artillery established a very heavy barrier, but all the aircraft came through and fulfilled their mission. *Adj/Chef.* Cuiller, from 4/13, was nevertheless seriously wounded and had to land in French lines, while the other crews returned to Meaux in their riddled aircraft.

This could be a success for the unit, but the men suddenly understood during this baptism of fire that they were powerless in front of the formidable enemy military machine. Their attack looked very much like the charge of the cavalry squadrons in Reichshoffen, and defeat already comes to mind. Moreover, the other units have suffered severe losses, and six of their comrades have been killed in this attack.

On May 18th, at noon, three German Dornier Do17s flew over the airfield of Meaux-Villenoy at very low altitude and dropped their bomb load while strafing, but to no avail. In the afternoon a cover mission took place over the Soissons-Simes sector. A flight took off, manned by *Cne* Treillard, *Adj.* Séghi, *Sgt/Chef.* Post, *S/Lt.* Merle and *Adj.* Voirin/*Sgt.* Sannier, attacking a Heinkel He 111 north of Braine and forcing it into an emergency landing. The leader then landed to capture the enemy crew, and bring back some trophies to the squadron. At 16:00, a formation of seventeen enemy bombers came over the airfield again, this time at 2,000 m (6,562 ft), and bombed the installations and barracks, without success.

On May 19th, new cover missions were carried out over the sectors of Creil, Château-Thierry, Fismes and Soissons. In the morning, a flight crewed by *S/Lt.* Merle, *Sgt.* Sannier and *Sgt/Chef.* Post, *Adj.* Boutillon, met north of Braine a formation of twelve Heinkel He 111s and attacked it, but their manoeuvre was handicapped by the French AA and the enemy bombers scattered into the clouds. A little later, they found one of the straying He 111s, which was immediately attacked from both sides and dropped its bombs before diving away trailing thick smoke. It was found the day after, shot down.

In the evening, a flight of two aircraft from Squadron 4/13, accompanied by the crew of *Sgt/Chef.* Post, *Adj.* Boutillon, intercepted a formation of six He 111s flying at high level between Compiègne and Soissons.

The three Potez concentrated their fire on the last bomber on the right, and forced it to drop from the formation and dive away with its port engine in flames. The other aircraft were too fast and were not reached.

The same missions continued until May 26th. On the 21st at 11:00, a flight made up of *S/Lt.* Boursain, *Sgt.* Jobard and *Sgt/Chef.* Post, *Adj.* Boutillon performed a cover mission over Creil and dived to attack a Do17 in collaboration with Bloch 152 single-seat fighters. The bomber immediately crashed in an open area in the forest of Halette, and its wounded crew was captured by ground troops.

On the night of May 21st, enemy bombers were streaming over the airfield and dropped some twenty bombs, without results. Their attack coincided with the suppression of the night fighter missions, which the unit could no longer sustain. On May 22nd, two Potez 631s fitted with a reinforced armament (two guns in the fuselage and four machine-guns under the wings) were added to the squadron's fleet, which now amounted to thirteen aircraft.

On May 25th at 17:30, a formation of thirty-six enemy bombers, escorted by Bf 110s, attacked the airfield and its neighbourhood, and dropped five hundred HE bombs of all sizes, plus one thousand incendiary bombs. During several minutes, the bombing was terrible, and the airfield disappeared under the smoke of the explosions. But the damage was of little importance, considering the means used - two aircraft sustained hits that will involve long repairs, and three others were slightly damaged; an armoury trailer and a fuel reservoir were destroyed; the telephone lines were cut; the tarmac was riddled with holes, which will rapidly be repaired by the maintenance company. Unfortunately, there were also several victims - Master-worker Doisy was killed, *Sgt.* Antoine was seriously wounded and had one leg amputated, and *Sgt/Chef.* Olivier and Sol. Dagneau were slightly wounded by shrapnel.

The alert had come too late and the enemy had taken advantage of the surprise, as only *Cne* Teillard was able to take off under the bombs. But he was powerless in the face of such numbers, and could have no influence on the events.

The permanent threat of enemy parachutists caused security to be doubled. Two light tanks were assigned to the airfield's defence, in such an event. But considering their slow speed (6 to 10 km/h), the command preferred to give this task to watch patrols.

The airfield of Meaux-Villenoy was full with many French units in retreat (*Groupe de Chasse* I/3 and its Dewoitine D 520s, *Escadrille de Chasse* 1/16 and its Potez 631s) as well as a British Spitfire reconnaissance unit. 1/13 must soon

This was Nîmes in 1941, and the Potez 631 C3s of ECN 1/13 were sporting the red and yellow Vichy markings.

SHAA via Autor

change base and move to an emergency strip surrounded by trees, at Trilbardou, near Meaux, where it stayed for one day and organized its camouflage.

On May 26, the unit moved again, to Melun-Villaroche, and on May 28th it settled at Moissy-Cramayel, nearby Brie-Comte Robert, where it will stay until June 11th. Between May 29th and June 4th, individual crews performed early warning missions, given the task to replace the failing ground listening systems, in order to signal to the rear lines, by radio, the arrival of enemy formations. Carried out for two hours at a time, these missions were particularly tiring and dangerous. The Potez 631s were frequently engaged by German fighters, and the crews suffered huge losses in this glory-less, boring task that will remain in everyone's memory as a real obsession.

On May 30, the crew of *Sgt/Chef.* Post, *Adj.* Boutillon and *Sgt/Chef.* Giloux was twice fired at by ground troops concentrated in the Compiègne forest, and the aircraft was hit twice in the wing. A little later, they came across a formation of eight Bf 109s that went by without noticing their presence.

On June 3rd, the crew of *S/Lt.* Giger, *Adj/Chef* Gobert, *Sgt.* Duffour did not come back from a scouting mission on the Lassigny-Vic sur Aisne axis. It beacme known the day after that the aircraft had been shot down by enemy fighters, and crashed in flames at Gournay-sur-Aronde.

On the same day, the crew of *S/Lt.* Boursain, *Sgt.* Jobard, *Sgt.* Dobsik was attacked by five Bf 109s, but escaped from them after a long low level pursuit. Still the same day, the crew of *S/Lt.* Merle, *Sgt.* Sannier, *Sgt.* Pelletier was fired at from the ground nearby La Frête-sous-Jouarre, and then attacked by a Bloch 152 fighter, and the aircraft was hit by two bullets. A little later, they met a flight of Bf 109s, that kept their heading without attacking.

On June 5th, *Cne.* Pelleray was assigned to *Escadrille* 2/13, in order to substitute for *Cne.* Petit de Nirbeck, killed in action. 1/13 had only four pilots left, including one NCO. From then on, the cover missions were resumed over distinct sectors, and all the regrouped squadrons performed together. For the first time, the night fighter crews were gathered in big and dangerous formations, but luck was not there, and they did not succeed in meeting the enemy to engage it.

On June 11th, the units flew to Saumeray (Eure et Loire), which was to become the base of all the gathered night fighter units, reinforced by Squadron 1/16. The lack of pilots compelled the latter to shuttle between both airfields, in order to convey the available aircraft. 1/13 recovered this way its eight Potez 631s. On the 12th, three enemy aircraft arrived by surprise from the clouds and bombed the surroundings of the airfield, on which were present some fifty aircraft, but they caused no harm. The fighters were surprised and had no time to enter into action.

The systematic cover missions to the east of Paris continue until June 13th. Then, the German advance towards Chartres was threatening the base of Saumeray, and the units were ordered to go to the airfield at Fontenay-le-Conte (Vendée). Four aircraft from the unit took to the air at 19:00. On their way, they were fired upon by French AA batteries, but not hit, and they landed at Fontenay-le-Comte at dusk.

On the following day, June 14th, a shuttle was organized in order to recover the aircraft still at Saumeray. The pilots took off, four per aircraft, and got the eight remaining planes back in two flights. During one of the flights, *Sgt.* Favard, a reserve pilot, was found at Tours and integrated into the squadron. His experience was insufficient to perform missions, but he would nevertheless be useful for the transfers, and fulfilled his task with courage.

On June 15th, all the units moved to La Saussaie (Charente Maritime). Orders arrived to continue with operational missions, but they cannot be obeyed, as there was no fuel. On June 17th, following an order from the General who commanded the south-east army, the units had to move to Les Milles, near Aix-en-Provence (Bouches du Rhône). Five planes from 1/13 took off, including one at the hands of *Sgt* Favard, but the others unfortunately had to be destroyed on the spot, not to let them fall into enemy hands.

On June 19th, several missions were planned by the high command, particularly regarding the protection of a bombing mission to Italy and the air cover of southern cities. But they will not be fulfilled, due to adverse weather conditions. An order arrived then to prepare the aircraft for transfer to Corsica and Tunisia. Finally, this solution was abandoned because of the problems it entailed, and the units stayed in France as a rearguard system, while almost all the air force crossed the sea.

On June 20th, the flying part of the unit left Les Milles and flew to Nîmes-Garons (Gard). The ground echelon followed and rejoined the unit - out from la Rochelle on June 17, it had gone through many escapades and crossed all France. During the following days, heavy rains soaked Garons airfield, a simple strip without any facilities, and the aircraft sank in the mud, a situation that forced the personnel to a temporary rest.

On June 23rd, as the armistice was required by *Maréchal* Pétain, the unit was ordered on a last mission, in order to repeat the attack of May 17th against German troops and armoured vehicles following the Isère valley up to their meeting with Italian troops, between Moirane and Grenoble. The state of the ground allowed only one aircraft to take off, crewed by *S/Lt.* Boursain, *Sgt.* Jobard. The other planes could not free themselves from the mud. On the other hand, the formations composed of aircraft of other units, able to take to the air, were considerably handicapped by a very low cloud base and had little opportunity to attack the enemy.

On June 25th, the war was over, as the armistice had been accepted. All aviation units were immediately ordered to stop flying. The engines remained quiet and the national colours on the ground. France was beaten, and those who had accepted it all, down to the last sacrifice, those pilots, mechanics, radio operators and others, were tasting the bitterness of defeat. But their conscience was clear - Squadron 1/13 had dutifully performed its task!

Between September 2nd, 1939, and June 25th, 1940, its crews performed, aboard aircraft that were inferior to the enemy's in number, speed, radius of action and armament:

- 33 night fighter missions
- 68 cover missions
- 7 strafing missions at very low altitude
- 28 anti-bomber scout missions
- 4 enemy aircraft had been shot down (confirmed)

All this was not in vain, as honour was intact. All through the citations bestowed on its crews, the squadron can read the witness of its efforts and sacrifices. On August 2nd, 1940, on the airfield of Nîmes-Courbessac, the Commander-in-Chief, *Gen.* Vuillemin hung the *Légion d'Honneur* on *S/Lt.* Boursain, and the *Médaille Militaire* on *Sgt/Chef.* Post.

There were others, another *Légion d'Honneur* (posthumous) for *S/Lt.* Giger, another two *Médailles Militaires* for *Adj/Chef* Gobert (posthumous) and *Sgt.* Antoine (wounded); a handful of *Croix de Guerre* went to *Cne.* Treillard, *S/Lts* Boursain, Merle and Giger (posthumous), *Sgt/Chef.* Post (with two citations), *Adj/Chef* Gobert (posthumous) and *Adj-s* Boutillon, Seghi, Voiron, Jobard, Sannier and Antoine. Cited at Fighter Command level: *Cne-s* Treillard and Pelleray, *S/Lt-s* Boursain, Merle and Giger, *Adj/Chef* Gobert, *Adj-s* Boutillon (two citations), Seghi, Saint-Aubert, *Sgt/Chef-s* Post, Lenoir, Blanc, Giloux, *Sgt-s* Jobard, Sannier, Dobsik, Pelletier and Cailleux; and at Division level, *S/Lt.* Boursain and *Sgt* Jobard.

Potez 63 Family versions

Potez 630 C3	–	fighter with Hispano-Suiza HS-14A/B engines
Potez 631 C3	–	fighter with Gnome-Rhône GR 14M4/5 engines
Potez 631 CN2	–	night fighter with Gnome-Rhône GR 14M4/5 engines
Potez 632 Bp2	–	naval dive bomber prototype
Potez 633 B2	–	light bomber, two seater
Potez 634	–	often used for training version **Potez 631 Ins**
Potez 635 CN2	–	night fighter prototype
Potez 637 A3	–	reconnaissance, three seater
Potez 639 AB2	–	attack, two seater prototype
Potez 63-11 A3	–	reconnaissance and army cooperation, three seater
Potez 63-12 C3	–	fighter with Pratt & Whitney Twin Wasp Junior engines, prototype
Potez 63-13 Bp2	–	dive-bomber prototype
Potez 63-16 T3	–	trainer prototype

Designation		English name
C3	–	Fighter, three seater
CN2	–	Night fighter, two seater
Bp2	–	Dive bomber, two seater
B2	–	light bomber, two seater
Ins	–	Trainer
A3	–	Reconnaissance, three seater
AB2	–	Attack, two seater
T3	–	trainer

The end of the French Potez 63

In April 1945, three Potez 631s (c/n 122, 134 and 185) from the former Vichy aviation were put back into service by the *Forces Aériennes de l'Atlantique*, which were charged with supporting the French troops fighting the Germans entrenched in the Atlantic bulges of Royan, Pointe de Grave and Oléron. They were integrated into GR 3/33 *Périgord*, equipped with the Fieseler Fi 156C *Storch*, and they performed reconnaissance missions over Oléron island at the end of the war, in May 1945. These were visual reconnaissance flights, as they were totally devoid of any photographic equipment. Once the war was over, they ended their careers at the CEV of Brétigny, used for the instruction of new *Armée de l'Air* pilots until 1947.

Foreign Potez 63s

The Potez 63 drew the attention of several foreign military forces, as it was, in 1938, a state-of-the-art aircraft. The Czech company Avia bought the production licence for a three-seat fighter variant with Walter engines, called Potez 636, but the invasion of Czechoslovakia by Germany put an end to this project.

China signed a contract for the acquisition of 4 Potez 631s and 5 Potez 633s. These aircraft were built, but as France went to war they were embargoed in Indochina just before being delivered to China, and assigned to the *Armée de l'Air*. Two of them were briefly used in September, 1940, during the Japanese coup at Lang Son. They performed reconnaissance flights over the Japanese columns that advanced through Indochinese territory. Another of these ex-Chinese Potez 631s was used in October 1940 as a command aircraft for GAM 42, based at Siem Reap during the Thai attack against the south of Indochina.

Yugoslavia ordered 2 Potez 630s and 1 Potez 631 for evaluation, and the first two were delivered, but the third was destroyed in France during tests.

Switzerland acquired in 1938 one Potez 632 and one Potez 630. The former was a variant of the Potez 633, but with Hispano engines that had been specially developed for Swiss service. It was registered HB-HAS with military code B-1. The Potez 630 was taken from the production destined for the *Armée de l'Air*, before being registered HB-HAT with military code B-2. These aircraft were evaluated for the needs of the Swiss Air Force, but though they proved to be satisfactory, France had enough problems providing its own equipment, and there were no further orders. The two aircraft flew with the *Troupe d'Aviation Suisse* between 1938 and 1944, one of them being fitted with two guns mounted in a removable belly gondola.

German forces also captured a number of French Potez 63s, and used some of them for training purposes.

A Yugoslav Potez 630 with a liaison De Havilland D.H.89 in the background.

The Swiss Potez 632, registered B-1, shortly after its delivery in 1938.

The same aircraft with the neutrality markings applied on Swiss aircraft after mid-September 1944. The belly gondola with two Oerlikon FF/K 20mm guns is clearly visible.

Potez 631 at scrap yard, Bretigny-sur-Orge. **Stratus coll.**

Potez 631 after overhaul by the Germans. **Stratus coll.**

A Potez 630 under Luftwaffe *colours in occupied France, 1941.*

An engineless German Potez 63 on a French airfield.

TK

Greek Potez 633 B2s

Greece ordered 24 Potez 633s, but in September 1939 when the French authorities decide to embargo all aircraft in production for foreign customers, not all of them had been delivered. This was why the EVA (Greek Royal Air Force) had only one Potez 633 unit on October 28th, 1940, when the Italian forces attacked. The 31st Squadron was part of bomber command, and had eight

Potez 633 of the 31st Greek Royal Aviation Squadron in the early autumn of 1940.

operational Potez 633s based at Niamata (Larissa) under the orders of Captain G.Karnavias. The Greek Potez 633s were registered B221 to B231, B233 and B234, that is, 13 aircraft all in all. But at that time, four of them had been lost in various accidents, like the one on September 7th 1939, which killed both crew members.

This Potez 633 of the Greek Royal Aviation (EVA) had belly-landed and been photographed by the Germans after they invaded the country in spring 1941. *JCS*

The Greek bomber command then consisted of the 31st Squadron and its eight Potez 633s, the 32nd Squadron with 11 Bristol Blenheim Mk.IVs, and the 33rd Squadron with 11 Fairey Battles. All through the campaign, the Greek Potez 633s operated without rest. Though in marked inferiority to the Italian units, their crews showed a remarkable courage and, on the ground, the Italian troops were compelled to retreat by the Greek forces' counter-offensive.

The Greek Potez 633's career was full of incident. Late on November 1st, 1940, three aircraft were flying back to their base at Larissa, after having bombed ground targets north of Ioannina. Two of them were able to land on the strip under the light of searchlights, but as the third one was on final approach at the hands of *Cdt.* Kousigiannis, the unit's CO, the base commander ordered all lights extinguished because he feared an attack from the Italians. Kousigiannis had to abort and head for the airfield at Tatoi. When he arrived he was completely out of fuel and made a dead stick landing, hitting an Avro at the end of the runway. Fortunately he was only slightly injured.

Wreckage of French impressed Potez 633 G17 in 1940 of GBA II/51, shot down near Amiens on 30th May 1940 by a Bf109E from I/JG 76.

On November 15th the aircraft B233 was shot down by Greek AA guns, and *Sgt/Chef*. Spiridon Kovatzis was killed while 2/Lt. A.Vladoussis had to bail out.

On December 21st B230 was carrying out a reconnaissance mission over southern Albania for the Greek army. It was intercepted by Fiat G.50*bis* of the 154°*Gruppo CT* and shot down over Valona (Vlorë). Captain Giorgios Karnavias and Ioannis Papadakis took to their parachutes, but were never seen again.

On December 30th the Potez 633s left their detachment base at Niamata to return to Larissa, which had better facilities for the harsh winter.

Potez 633 c/n G17 (B-237) ex-Greek.

Another ex-Greek Potez 633 on a French airfield near a Potez 63-11.

On February 10th 1942, another aircraft was shot down over Albania by G.50*bis* of the 154° *Gruppo CT*, and its pilot, *Lt.* Giorgios Stravaetos, was killed.

April 6th marked the beginning of Operation Marita, as Germany came to reinforce its Italian allies struggling against the Greeks. The air superiority of the Axis forces was then overwhelming, in spite of reinforcements sent by the RAF. On April 15th two Potez 63s were written off on the ground during a raid against Larissa. It was the end for Greece, and after the occupation of the country by German forces, nobody knows what happened to the surviving Potez 63s, if indeed there were any.

Romanian Potez 633 B2s

Contracts for two batches of 20 Potez 633 B2s were issued by the Romanian government, and the first order was delivered in 1939. A new bomber group was set up for the new aircraft, Group 2, which included the 74th and 75th Squadrons and was integrated in the 2nd Bomber Wing.

As from June 22nd, 1941, the 2nd Wing was fully operational, under the command of Captain Ioan Cristescu-Cutitaru. The 74th Squadron was commanded by Captain Gheorghe D.Popescu, and the 75th by Captain Nicolae Bals. The group was transferred on June 19th from Buzau airfield to Sihiele, with an emergency airstrip at Boteni in case the base was attacked.

The two Squadrons of the 2nd Wing had the following operational machines:

74th Squadron:

- Flight 1: aircraft c/n 15 and 11, 18 and 12
- Flight 2: aircraft c/n 8, 9 and 1

75th Squadron:

- Flight 1: aircraft c/n 19, 4 and 13
- Flight 2: aircraft c/n 3, 17, 20 and 14

Potez 633 c/n R37 (X-756). Aircraft of the second batch ordered by Romania. This machine was left in France and was used against the Germans.

One of the Potez 633s ordered by Romania (order no 271/8 of April 26th 1938) but retained in France and used by the A d'l A. Potez 633 B2 c/n R31 (X-750) with Polish personnel at Caen. **BB coll.**

In the Stalingrad sector in deep winter, two airmen study their flight plan. MA

Romanian Potez 633 tactical n°12, still with the original French camouflage scheme, with the nickname "Sâc-Sâc" painted under the windscreen. MA

For a short time, the 4th Bomber Group and its PZL.37 *Łoś* were coupled with Group 2. When Romania went to war, both the 74th and 75th Squadrons were exclusively equipped with Potez 633s, but from July 16th the 75th received half a dozen Savoia JRS-79s, a modified variant of the Italian bomber, fitted with two Jumo engines and produced in Brasov by the IAR company.

Barbarossa: the Odessa Campaign

On June 21st, 1941, Group 2 was placed under alert, and at 20:00 Romania entered the war alongside Nazi Germany, ready to take part in the offensive launched by its ally against the USSR. On June 22nd at dawn, the aircraft of the 74th and 75th Squadrons were launched to attack Soviet airfields and positions. The first missions were marked by the loss of *Captain* Nicolae Bals, CO of the 75th, who was replaced by *Captain* Iuliu Potora.

In the daily missions, the Potez 633 was very much appreciated by its crews. The Soviet fighters were still dangerous, and the missions were carried out under escort by He 112, IAR 80 or Bf109 units. On July 4th the aircraft flown by *Sgt/Chef*. Stoica Craciun and *2/Lt*. Nicolae Grajdeanu did not return to base, and four other aircraft returned home damaged. Grajdeanu and Craciun were declared missing in action.

On July 28th, before the offensive against Odessa, the surviving Potez of the 75th Squadron were transferred to the 74th, the former unit being entirely re-equipped with Savoia JRS-79Bs. There were still five operational Potez 633s, plus one under repair. On August 20th another Potez was barely able to return home after having been hit by AA. On the 27th, while a flight of three Potez 633s was attacking a column over the Mariental-Alexanderhilf area they were intercepted by four Polikarpov I-16s. In addition to the fierce Soviet AA, this led to two of the three aircraft returning to base with serious damage.

During 1943 on the Russian front, Romanian Potez 633 crews ready themselves for a mission.
MA

Romanian Potez 633 on the Eastern Front in 1943. The upper part of the bomb bay is clearly visible under the canopy.

MA

On August 28th, several attacks were launched against targets in the Katarsov-Danik-Mariental area, and one Potez 633 did not return to base. The crew of *Sgts* D.Tudor and Alex Simulescu were declared missing in action.

On September 7th, Group 2's equipment was reinforced in view of the offensive against Odessa. The unit now included Potez 633s, Savoia JRS-79Bs and PZL.37 *Łoś*. On the 7th and 8th, several aircraft returned with damage after combats against I-16s, and on the 12th the Potez 633 crewed by Lts Constantin Popa and Stefan Pagu was shot down. On September 17th, the group only had some PZL.37s in flying condition. Finally, on October 16th, the last missions of the Odessa campaign took place.

During the Odessa campaign, a formation of Potez 633s escorted by He 112s.

MA

On the Eastern Front

November 1942 on the Stalingrad front. A camera is about to be fitted to a Potez 633 of a 1st Long Range Reconnaissance Group's squadron.

MA

Romania operated on the Eastern Front during the whole war. After the fall of Odessa, what was left of the 2nd Wing settled in Tecuci to be re-equipped. Nine Potez 633s from the *Armée de l'Air d'Armistice* were then delivered by Germany by September 25th 1941 in order to make up for attrition. A new Combat Air Group, or GAL, was formed with fighter, reconnaissance and bomber units. Its task consisted of ensuring the air cover of the Eastern Front portion assigned by Germany to the Romanian forces, in the Don and Stalingrad sectors.

In September 1942, the 1st Long Range Reconnaissance Group arrived in Stalingrad with the following units:

- 1st Long Range Reconnaissance Squadron, with four Potez 633s and six Bristol Blenheims
- 2nd Long Range Reconnaissance Squadron, flying Dornier Do 17s
- 3rd Long Range Reconnaissance Squadron, with four Potez 633s

In early December 1942, after an intensive operational period from rudimentary strips and under extreme weather conditions and hard combat, the Potez 633s and Do 17s were sent back to Romania for overhaul. They were eventually retired from operational flying and used for training.

At the same time, a 3rd Light Bomber Group had been created with three squadrons:

- 73rd Light Bomber Squadron, flying PZL23 *Karaś*
- 74th Light Bomber Squadron, with Potez 633 and 63-11s
- 81st Light Bomber Squadron, flying IAR 37s

These units took an intensive part in the combats over the Don and Stalingrad, during which two Potez 633s were lost. By the spring of 1943 the aircraft were war-weary, and the Potez 633s were withdrawn from the front. The 3rd Light Bomber Group was then re-equipped with Ju 87 D-3 *Stukas*, to be renamed 3rd Dive Bomber Group.

The surviving Potez 633s were overhauled and delivered to the 1st Fighter Wing, which used them for night flying instruction.

Detail photos 630-637
Fuselage

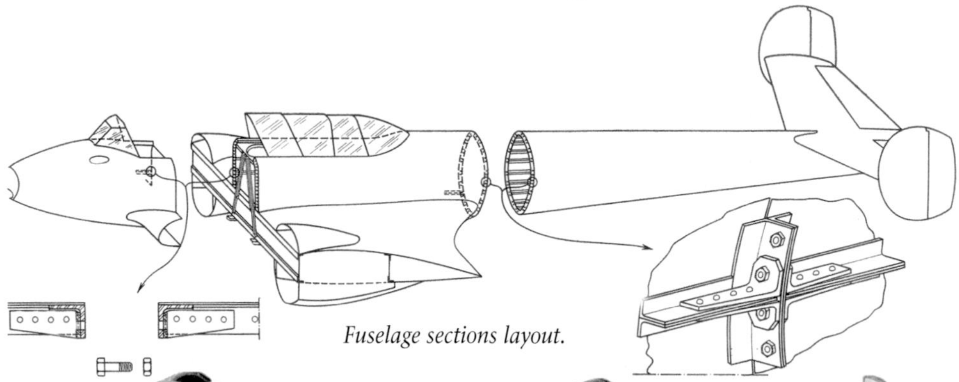

Fuselage sections layout.

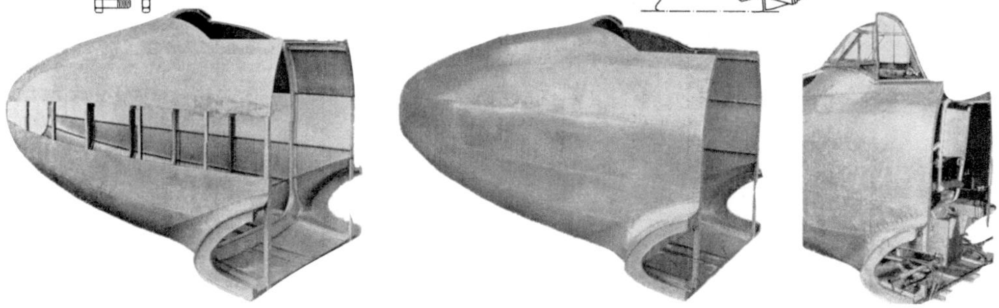

Potez 630 nose before its assembly on a fuselage.

Nose of Potez 639 with downward-tilting 20 mm cannon.

Rear fuselage sections.

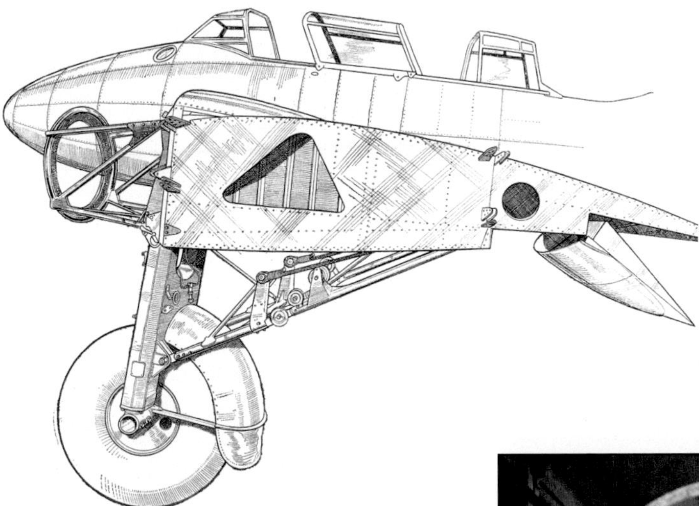

This drawing is from the French magazine l'AERONAUTIQUE (December 1938). It shows the engine bearers and the undercarriage.

Fitting cockpit accessories.

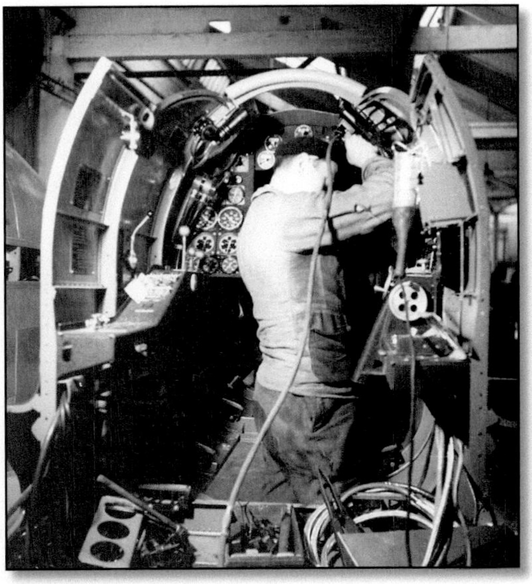

Cockpit

Below: The pilot's instrument panel.

Above: Canopy open and closed.

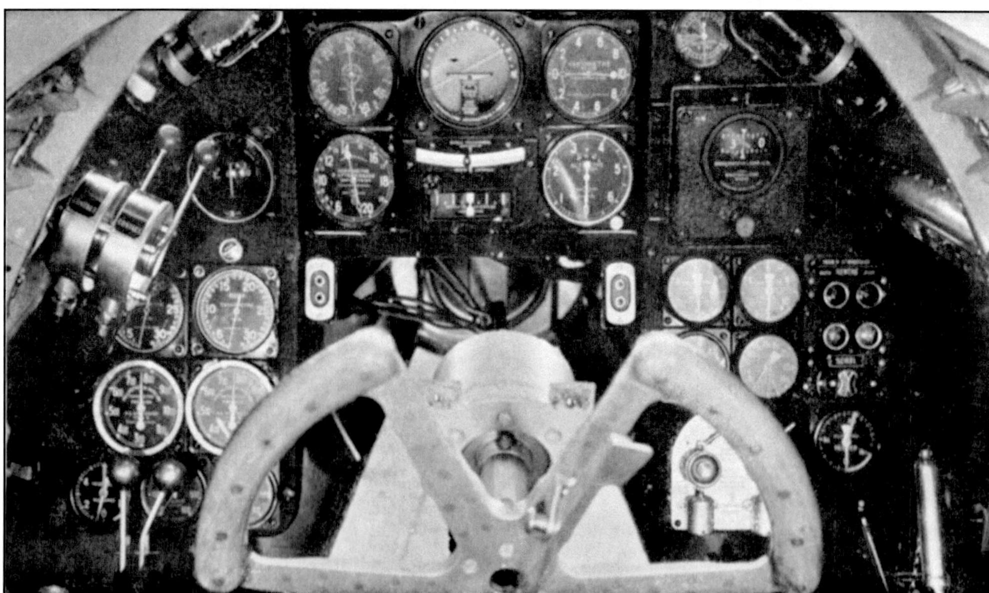

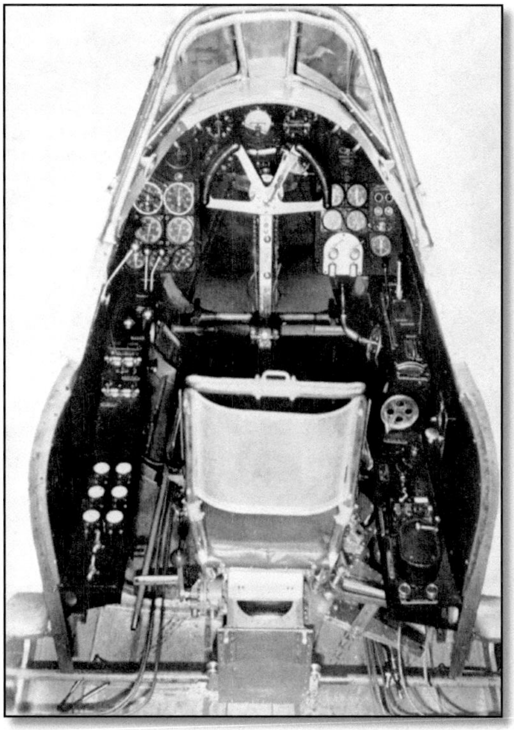

Above left: *Pilot's cockpit from above.*
Above right: *The cockpit seen from aft.*

Left: *The pilot's right-hand side controls.*

Below: *The pilot's left-hand side controls.*

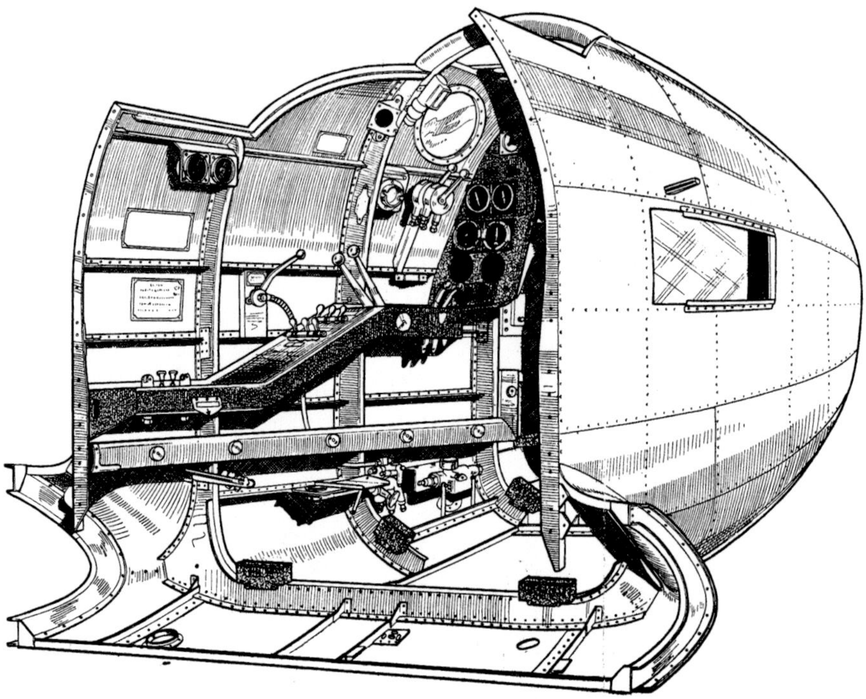

Another drawing from the French magazine l'AERONAUTIQUE (December 1938).
***Top:** the nose of the fuselage inside which we can see the left side of the cockpit.*
***Bottom:** we see here the mounting stand of the nose of the fuselage.*

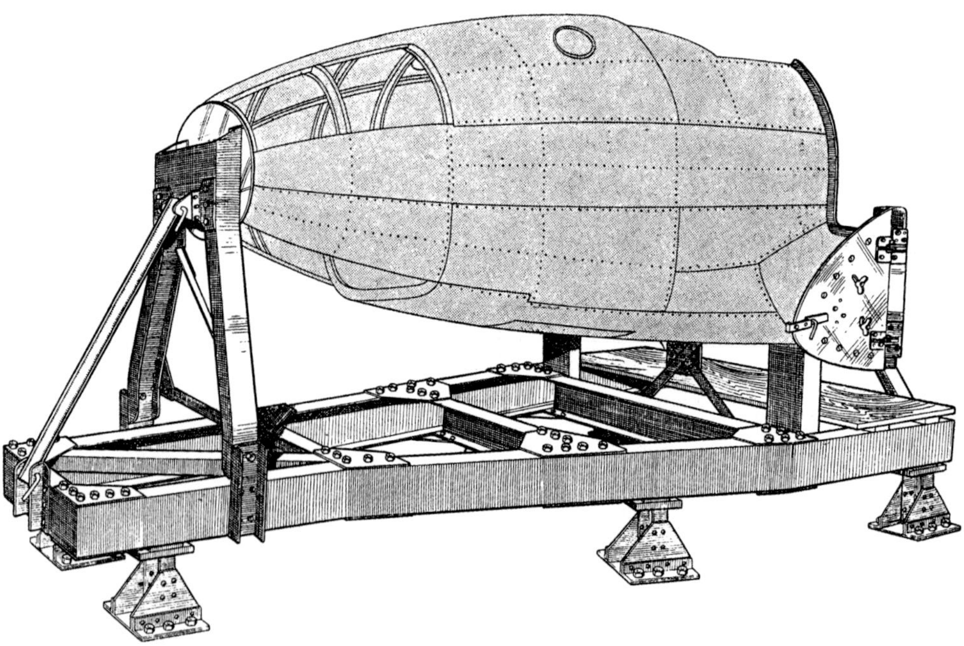

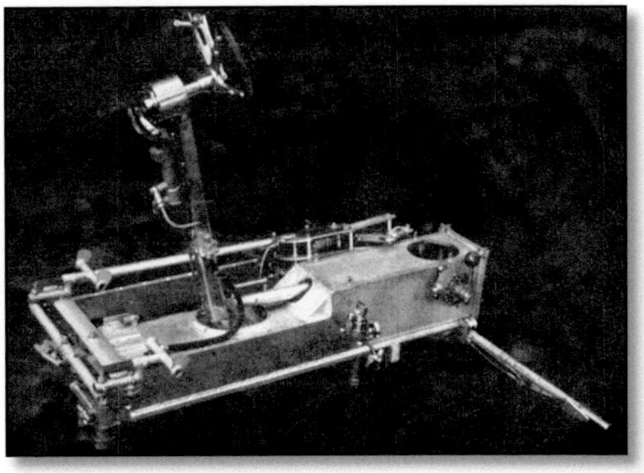

Left: *One-piece control set, ready to be installed in the fuselage.*

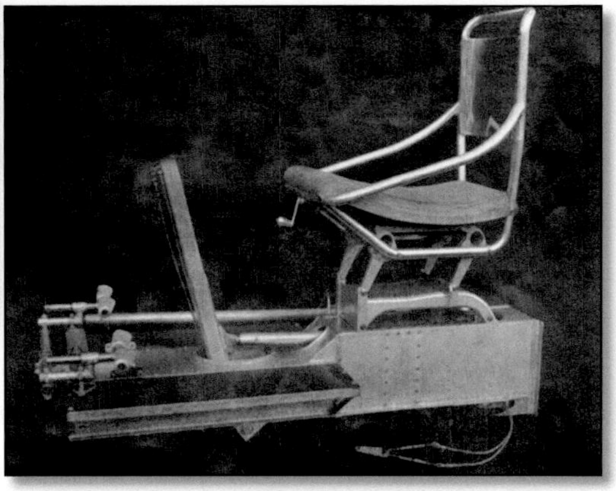

Right: *Cockpit floor with seat.*

Below: *Central section of fuselage.*

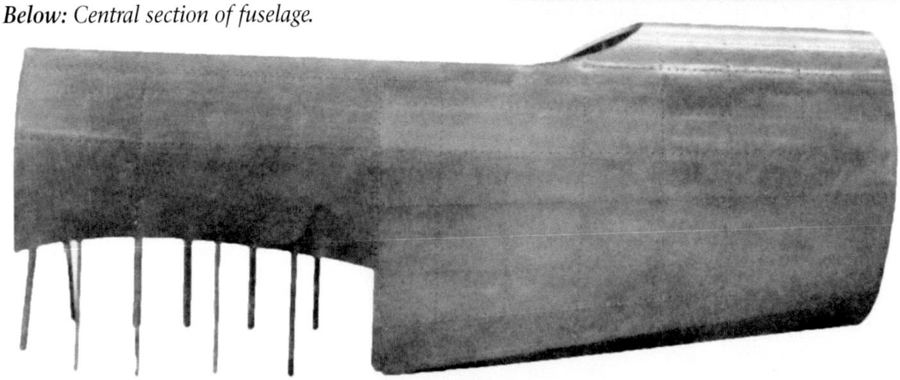

Fuselage and wing central section.

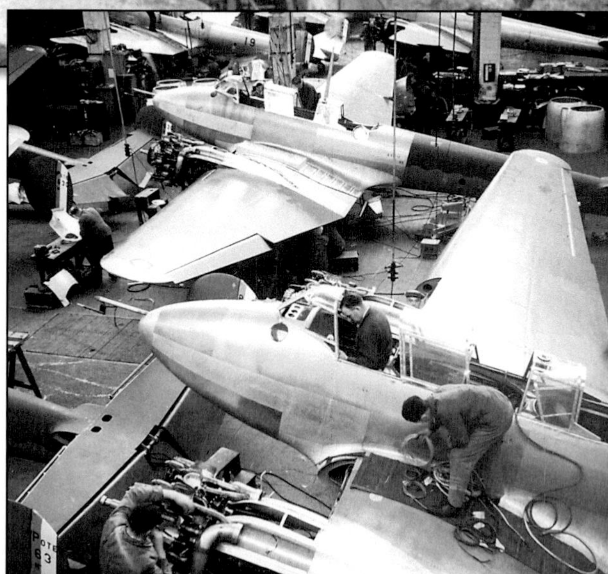

Top left and right: *Detail of a reconnaissance Potez 637, back from a mission over the German lines. It had been hit by Flak, and one can see the impact of a shell on the left side of the gondola, plus shrapnel impacts on the right side. The observer was probably killed or seriously wounded.*

Middle: *This Potez 637 clearly shows its belly gondola.*

Bottom: *Meaulte plant assembly line for the Potez 631.*

Armament

Above: Potez 631 c/n 74 during the mission

Left: The rear gunner of a night fighter Potez 631, and the 7.5mm MAC machine-gun. This picture may have been taken at Le Plessis-Belleville in March 1940.

Cannon 20 mm Hispano Suiza HS404.

Potez 630 c/n 18 (C-530) with a Potez 63-11 and a LeO 20. BB coll.

Wings

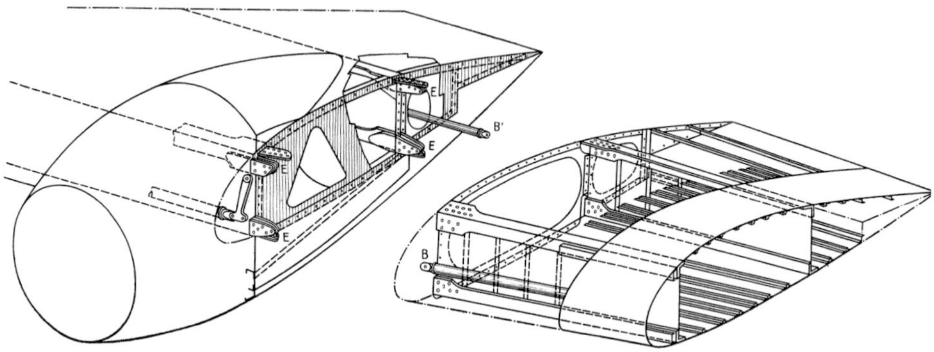

Outer wing section mounting points.

At work on outer wing panels.

Work on center section before fitting landing gear and outer wing panels.

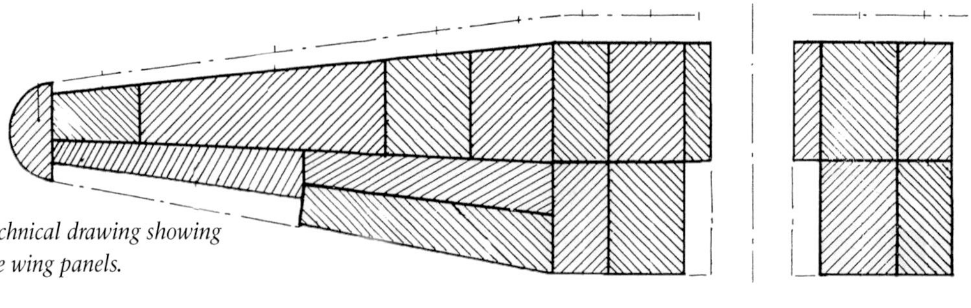

Technical drawing showing the wing panels.

Uncompleted starboard outer wing panel showing construction details.

Complete port outer wing panel of a Potez 63.

Potez 63 center sections under assembly.

On this Potez 631 (c/n 34?) one can see on the right wing the small roundel adopted at the beginning of the war.

A starboard outer wing panel with flap missing and aileron uncovered.

A starboard outer wing panel without undersurface skin.

Potez 630 seen from behind.

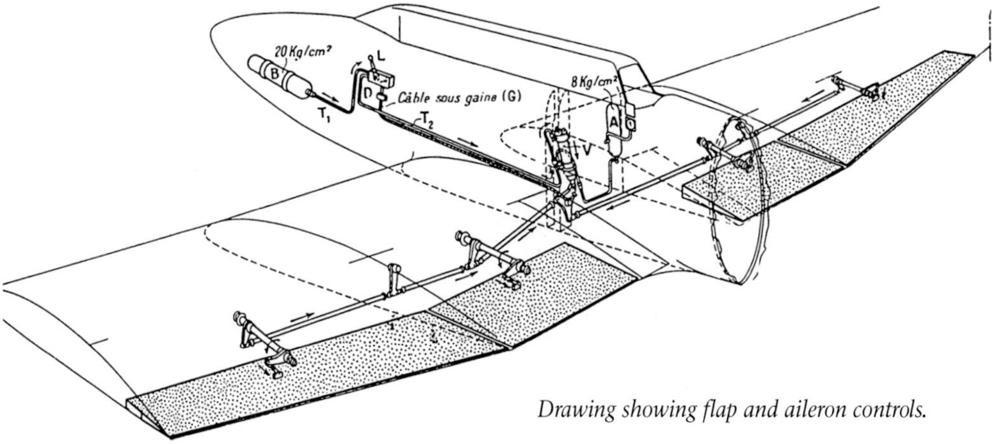

Drawing showing flap and aileron controls.

Aileron and flap construction

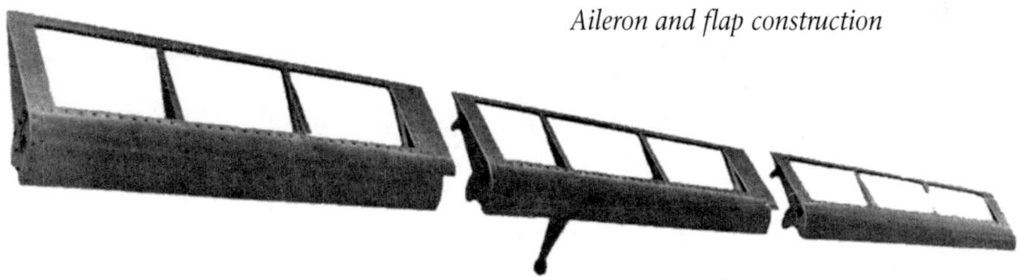

Wing and flaps details.

Tailplane

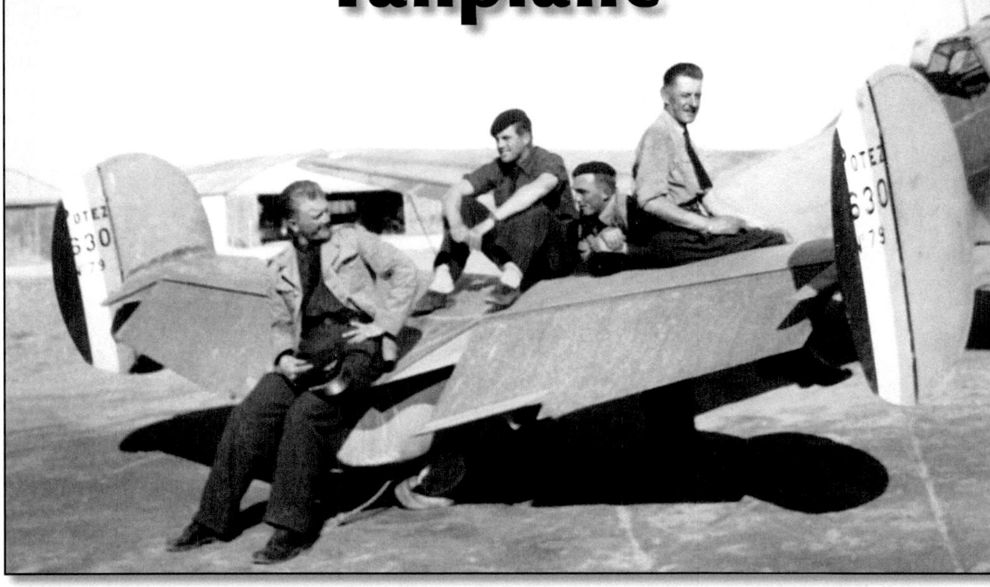

Tailplane of Potez 630 c/n 79. *JA via JR*

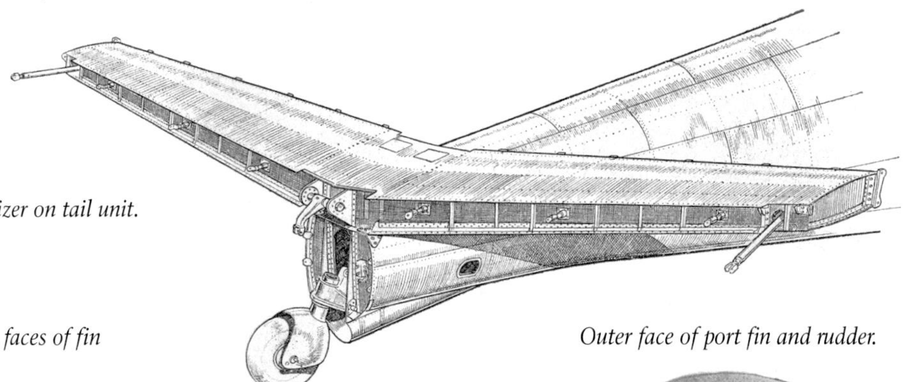

Stabilizer on tail unit.

Inside faces of fin

Outer face of port fin and rudder.

Above: *Right stabilizer tip and fairing with the vertical surface.*

Right: *Closeup view of Potez 630 c/n 02 tail. In background Potez 540.*

Below right: *Potez 631 c/n 135 from 1. Escadrille GC I/2, based in Chartres in August 1939. Above the stork insignia was written "Anne-Marie".*

Above: *Examination of F-AREY's tailplane by Yugoslav aviators.*

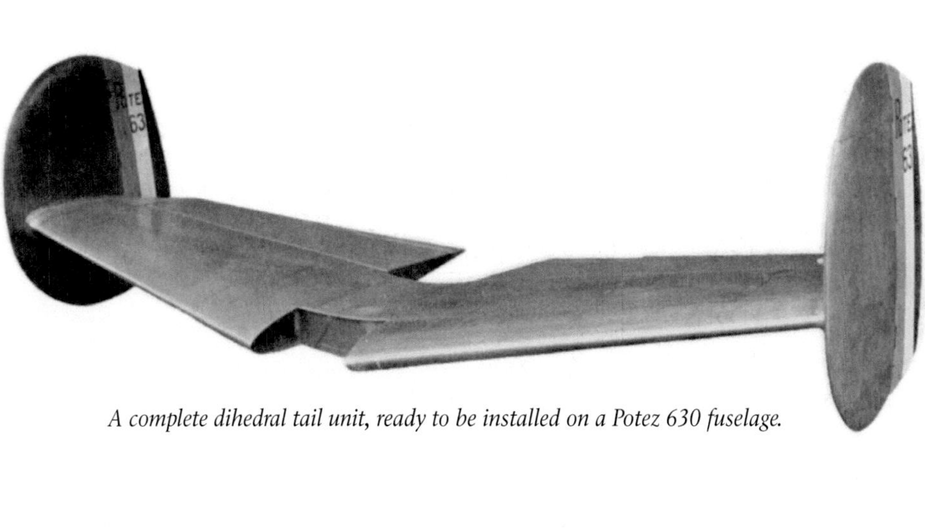

A complete dihedral tail unit, ready to be installed on a Potez 630 fuselage.

Engine

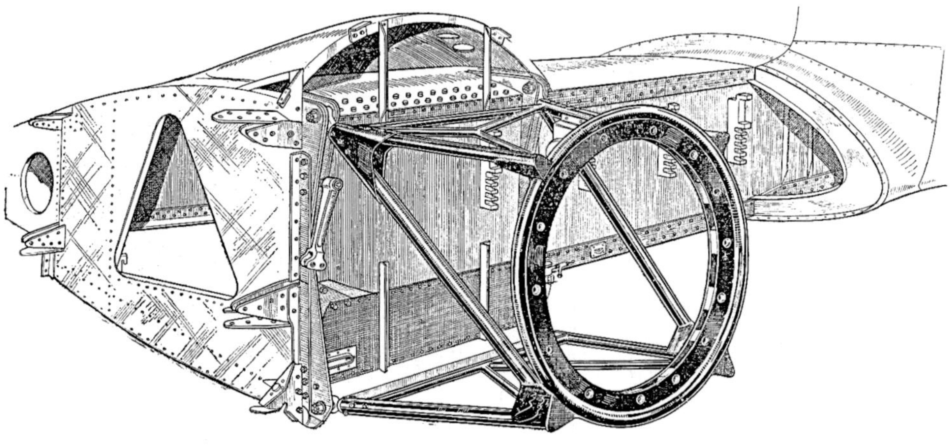

Engine mount.

This Potez 631 shows its Gnôme-Rhône 14M4/5 engines.

Engine nacelles and cowls of Potez 630.

Undercarriage

The tailwheel.

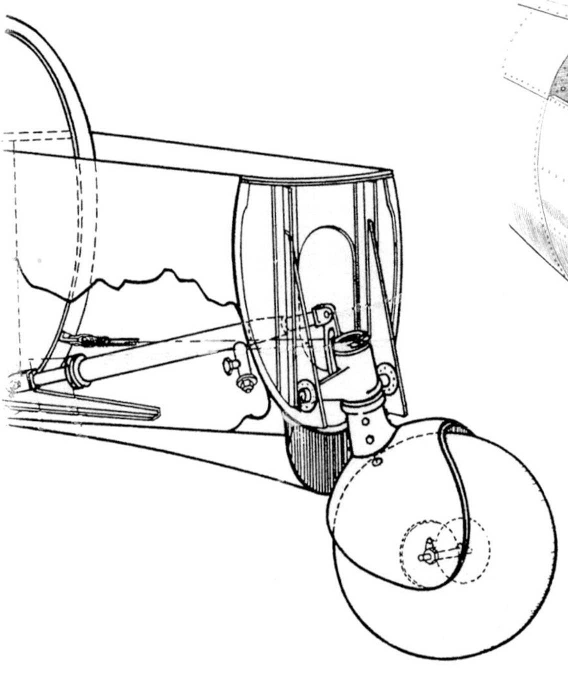

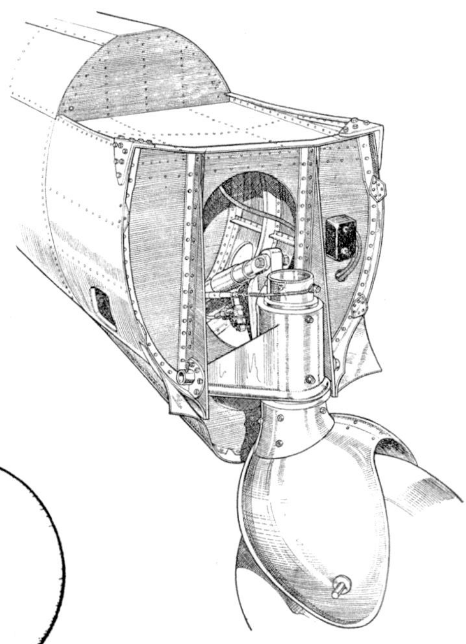

Tailwheel drawings.

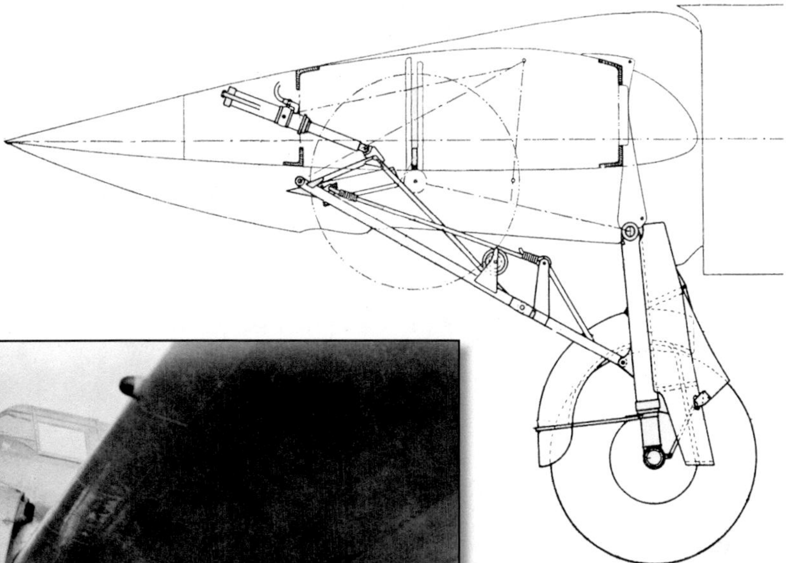

Wheel retraction scheme

Engine nacelle and landing gear of Potez 631.

This Potez plant official picture shows the details of a Potez 63 undercarriage.

Landing gear.

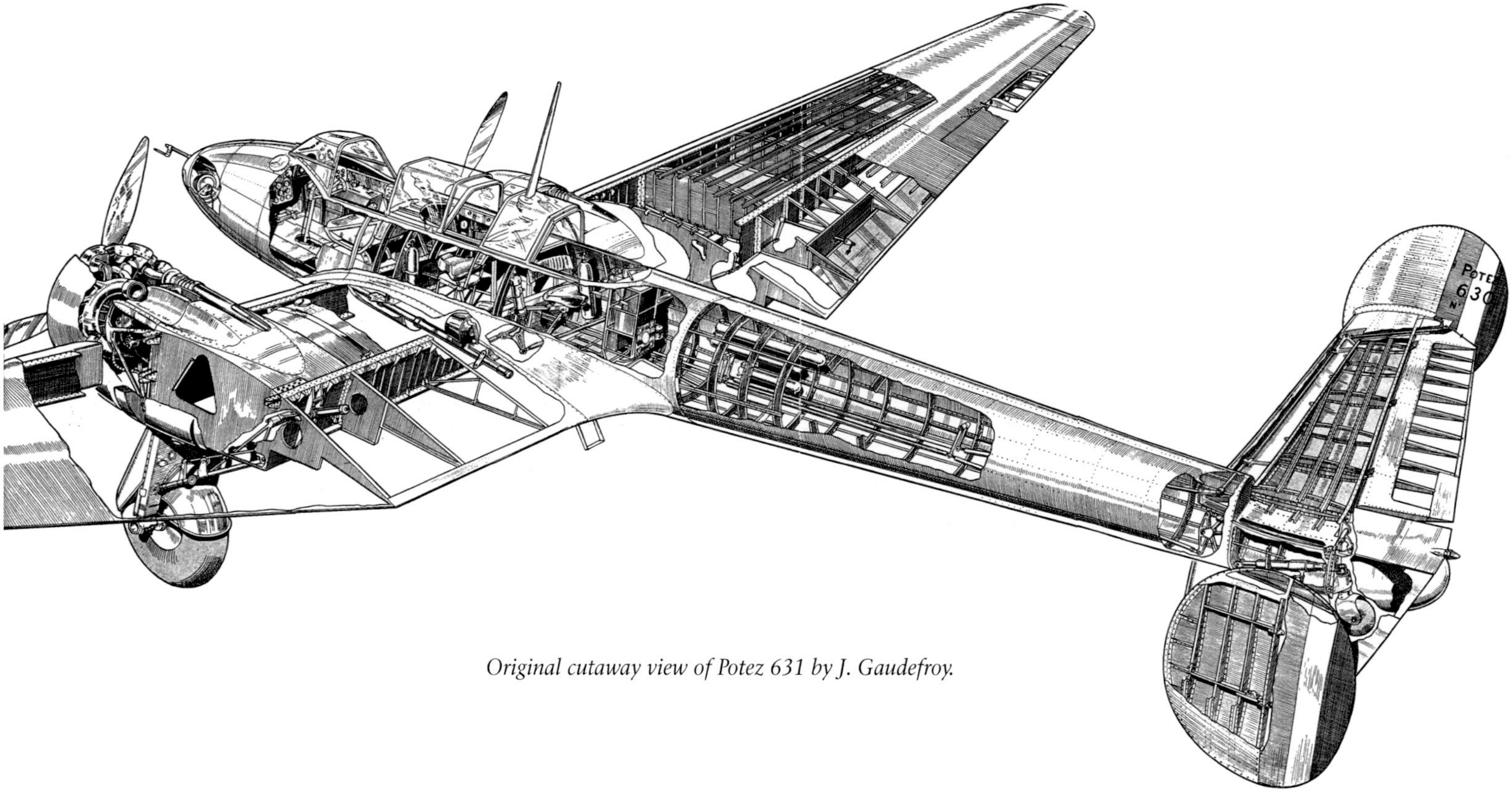

Original cutaway view of Potez 631 by J. Gaudefroy.

Potez 63-11

A brand new observation and reconnaissance aircraft was developed on the basis of the Potez 63 airframe. It was first destined to replace the outdated single engined aircraft still in service in observation groups, such as the Potez 25, Mureaux 115/117 and Breguet 27. The new Potez soon also replaced the Potez 542 and Bloch 131 in reconnaissance units, and the Potez 637s which were nothing more than an temporary expedient.

A Potez 631 was modified to become the first 63-11. The cockpit was redesigned, and the nose was entirely new - a large, glazed area was home for the observer, who enjoyed exceptional room and visibility. This was a far cry from the Potez 637, in which the observer was squeezed between the pilot and the gunner in a narrow space, and with a visibility much limited by the glazed panels of the belly gondola. The 63-11 prototype made its first flight on December 31st 1938, and subsequently an order for 415 aircraft was signed. The production aircraft had a modified, more streamlined clear nose with better vision. The rounded forms of the prototype's nose were found to handicap the observer's vision, and the new nose was consequently made of flat Plexiglas panels.

A test crew in front of the Potez 63-11 prototype.

Potez 63-11 prototype, with rounded nose.

Potez 63-11 c/n 1 at Villacoublay.

PRODUCTION

As with other Potez 63s, the 63-11 suffered from the slow production of Ratier three-bladed propellers, which lagged behind the pace of aircraft production. They were thus delivered with wooden two-bladed propellers, which lowered the maximum speed by some 40 km/h (25 mph). This was not the only problem affecting Potez 63-11 production, as the engines and armament did not follow pace either. Consequently in early September 1939, when France went to war, there were only five Potez 63-11s in the *Armée de l'Air*, and even these were not in operational service. With two Gnome-Rhône 14M engines driving Ratier 1527/8 or GR 2190/1 propellers, and fuel tanks for 950 litres (209 gallons), increased to 1120 litres (246 gallons) from January 1940, the aircraft had a 1,700 km (1,056 mile) range.

The Potez 63-11 assembly line, where several aircraft can be seen: c/n 616 in the centre, c/n 566 and c/n 567 in front of it.

Potez 63-11
1/72

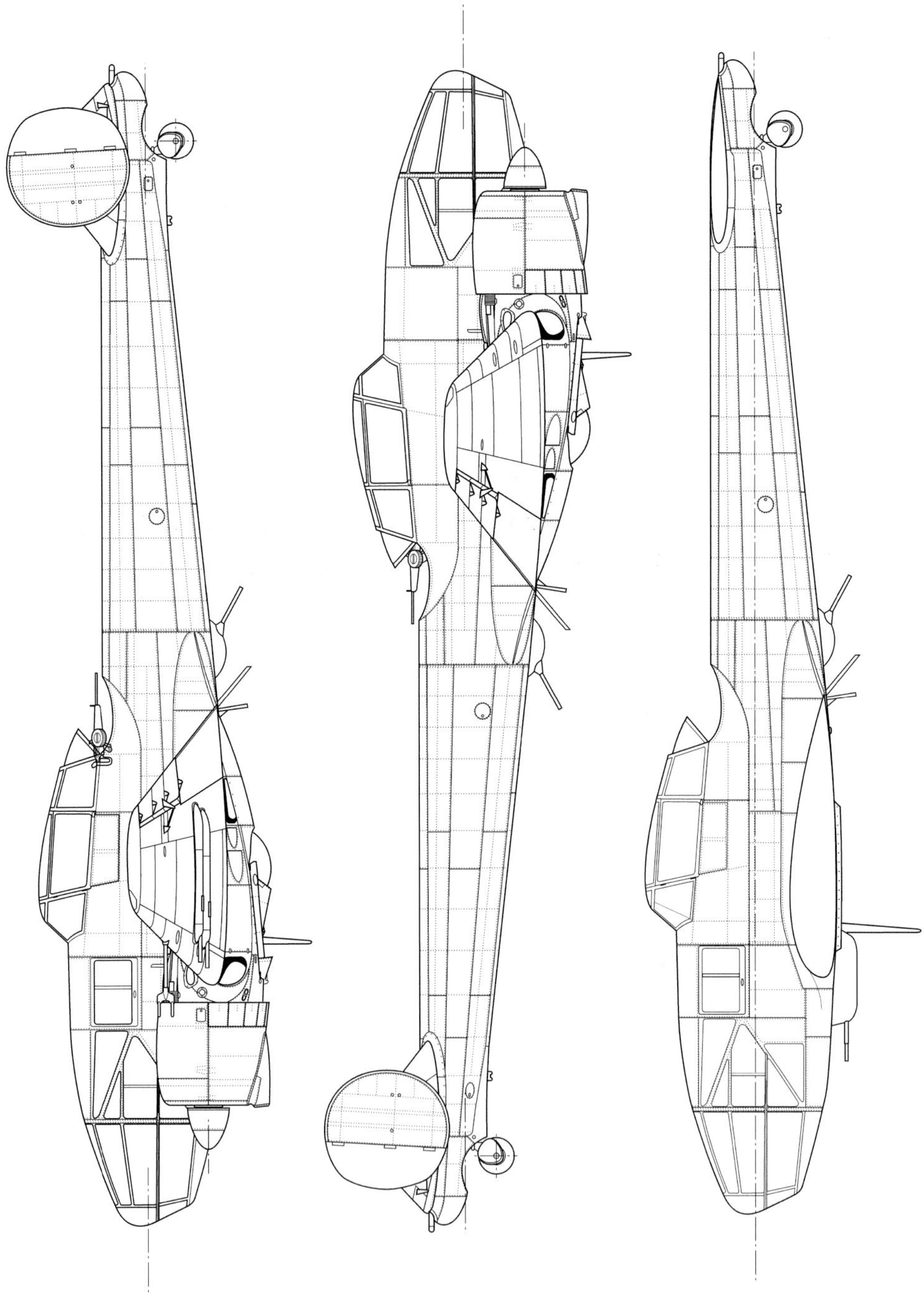

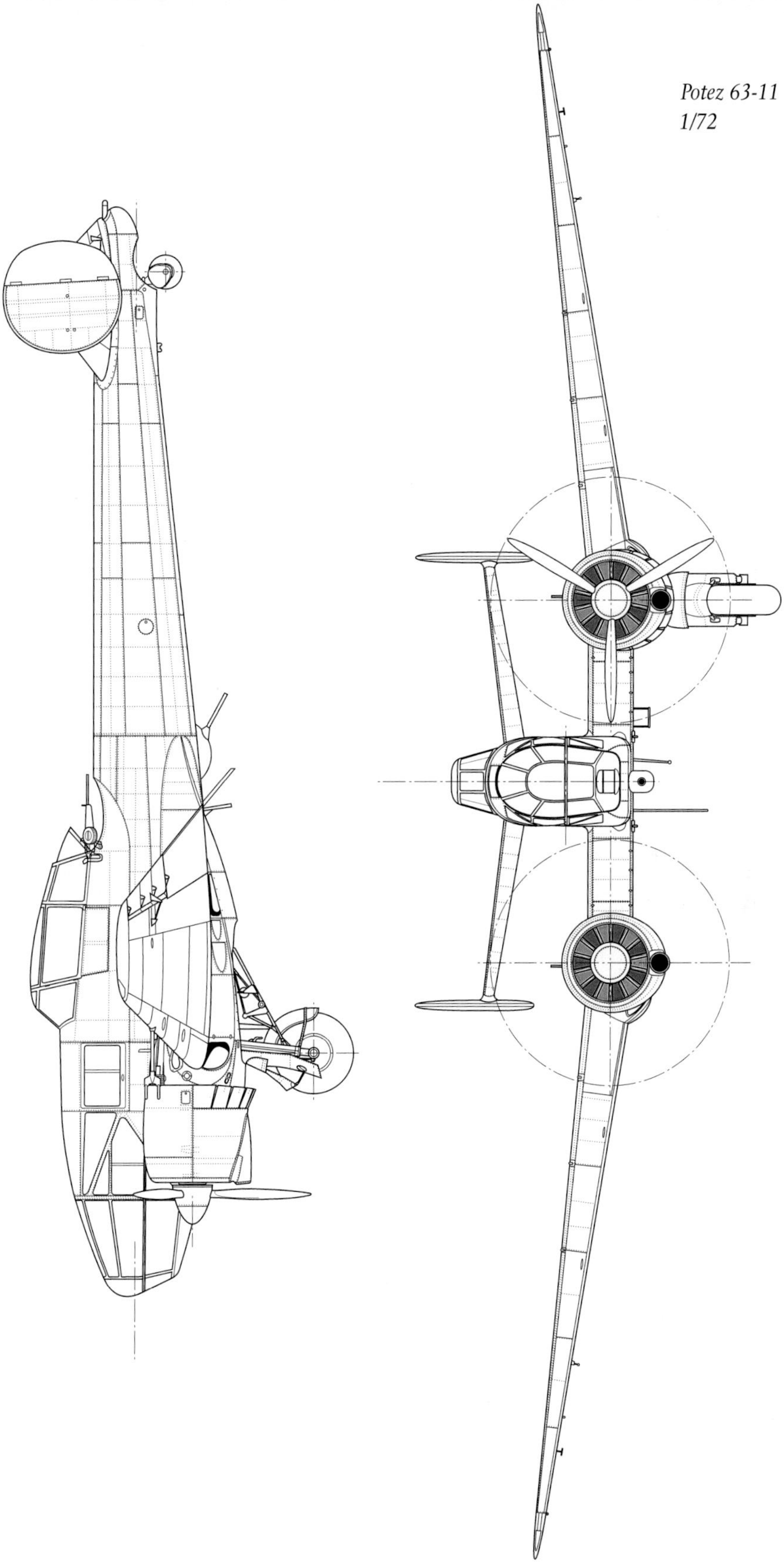

Potez 63-11
1/72

The armament initially included a fixed forward firing machine-gun and another fixed rear weapon firing downward. For defence, the gunner was given a MAC 34 on a dorsal flexible mount. During the first half of 1940, a number of the Potez 63-11s received reinforced armament. This comprised three forward-firing machine-guns under the fuselage and carriage for four 50 kg bombs, while some had an extra four machine guns under the wings. All this gave the aircraft a not inconsiderable ground attack capacity. The empty weight of 3,136 kg (6,194 lb) increased to 5,430 kg (11,971 lb) in operational conditions. With a practical ceiling of 8,500 m (27,887 ft), the maximum speed was 337 km/h (209 mph) at sea level and 425 km/h (264 mph) at 4,000 m (13,123 ft).

Potez 63-11 c/n 54 and 274 being cannibalized, with a background of four Caudron Simouns and two Spad 510s.

Uncompleted Potez 63-11s in the Méaulte plant taken over by the Germans.

Unknown Potez 63 -11, winter 1939/40.

Potez 63-11 c/n 598

Potez 63-11 c/n 366, in the markings of GR II/14, flying over Avignon on 1941.

In Service

According to Raymond Danel, Potez 63-11 deliveries on March 14th 1940, stood as follows: 459 aircraft accepted, including 217 with Gnome-Rhône 14M6/M7 engines and GR propellers; 95 aircraft with Gnome-Rhône M4/M5 engines and Levasseur propellers; 147 aircraft with Ratier two-bladed wooden propellers. Among these, 294 aircraft had been assigned to units, and 49 were in workshops, waiting for their armament to be reinforced.

The units receiving the Potez 63-11 were essentially the Reconnaissance Groups (109 aircraft) and the *Groupes Aériens d'Observation* (42 aircraft). Some 33 other aircraft were assigned to North African units, and 19 more to those in Syria.

Polish personel in Clermont-Ferrand during inspection of French Air Force command. Spring 1940.

AM

In spite of the delays in the delivery of the 63-11, production aircraft reached units in relatively large quantities, before other much needed types, like the Bloch MB 174, the Amiot 351 and 354, and the imported Martin 167 Maryland.

GR II/33, previously equipped with the Potez 637, was the first unit to take delivery of the 63-11 (c/n 45, 153, 154, 156, 165 and 189), to make up for attrition and aircraft withdrawn from service. The new plane was not particularly warmly welcomed. The observers were happy, as they were much more comfortable to fulfil their mission, but the pilots were less enthusiastic; they found the aircraft had lost manoeuvrability and speed when compared to its predecessor. The reconnaissance groups were attached to different armies, and worked for the latter's headquarters. They received six Potez 63-11s each:

- for the VIIème *Armée*, GR I/35 at St Omer-Wizerne
- for the Ière *Armée*, GR I/14 at Saint-Simon-Clastres
- for the IIème *Armée*, GR II/22 at Chatel-Chehéry
- for the IIIème *Armée*, GR I/22 at Metz-Frescaty
- for the IVème *Armée*, I/36 at Martigny-lès-Gerbonveaux
- for the Vème *Armée*, II/36 at Neufchâteau-Azelot
- for the VIIIème *Armée*, I/55 at Lure-Malbouhans
- for the *Armée des Alpes*, II/14 at Valence
- for the VIème (Reserve) *Armée*, II/55 at Chambarand-Marcilloles

Above: *Overall view of the parade square in front of the hangar, where troops were inspected, reports made, and flags raised ceremonially.*
Right: *March 27th 1940 - Lyon Bron. Official ceremony with gen. Sikorski.*
PI&SM coll.

Below: *French instructors with Polish pupil on St-Jacques - Rennes base.*
JS

Potez 63-11 c/n 287 of GR II/14.

Potez 63-11 c/n 211 of GR I/52.

Potez 63-11 and a Curtiss H-75.

Nose of Potez 63-11 c/n 296.

As early as December 1939, the old single engine aircraft equipping the GAOs were progressively replaced by Potez 63-11s. Twelve of these groups receive 9 brand new aircraft each, while 26 squadrons of other units received 6 each. They carried on their observation missions for the benefit of the army corps and the big mechanized units to which they were assigned:

- VIIème *Armée*: GAO 501 at Dunkerque-Mardyck, GAO 516 at Calais-St Inglevert, GAO 552 at St Omer-Wizerne.
- Ière *Armée*: GAO 503 at Valenciennes, GAO 504 at Denain-Prouvy, GAO 505 at Vertain-Le Quesnoy, GAO 502 at La Fère-Courges, GAO 544 at Villers-Lès-Guise
- IXème *Armée*: GAO 511 at Villers-Lès-Guise, GAO 547 at Mézières Tournes-Belval, GAO 2/551 at Tournes-Belval, GAO 545 at Denain-Prouvy, GAO 4/551 at Le Quesnoy
- IIème *Armée*: GAO 2/520 at Challerange, GAO 510 at Attigny, GAO 518 at Challerange, GAO 507 at Attigny
- IIIème *Armée*: GAO 2/506 at Chambley, GAO I/506 at Conflans-Doncourt, GAO II/508 at Mars-la-Tour, GAO I/551 at Etain-Buzy, GAO 3/551 at Senon-Spincourt

Potez 63-11s c/n 392 and 342

Potez 63-11 c/n 192 of GR I/14 shot down on May 15th 1940 by German Flak, close to Douai Dechy.
Stratus coll.

- IVème *Armée*: GAO 509 at Delme-Essey, GAO 520 at Morhange
- Vème *Armée*: GAO 548 at Epinal-Dogneville, GAO 512 at La Perthe, GAO 517 at Neufchâteau, GAO 553 at Nancy-Azelot, GAO 546 (reserve) at Sézanne
- VIIIème *Armée*: GAO 543 at Luxeuil-St Sauveur, GAO 513 at Belfort-Chaux
- *Armée des Alpes*: GAO I/584 at Valence, GAO 2/514 at St Etienne-de-St Geoirs,GAO I/514 at Montbard-Touillon
- In reserve: GAO 581 at Marignane, GAO 582 at Valence, GAO I/589 at Fayence-Sisteron, GAM 550 (observation squadron) at Ghisonaccia, in Corsica

In total, 34 GAOs with their full complement of six Potez 63-11s were operational, as were another six with a partial complement.

Burnt out remains of a Potez 63-11 at Tournes-Belval near Charleville-Mezieres, surrounded by fully serviceable Bf 109Es of Stab I./JG 1, a Gruppe that occupied this field between 17 and 22 May 1940.

Stratus coll.

The French Campaign

In May 1940, 238 aircraft were operational out of 396 delivered to reconnaissance units, a far cry from the 860 planes initially planned. These aircraft and their crews carried out a remarkable task, keeping headquarters informed of the enemy's moves day in and day out. These missions were very expensive, in aircraft and in lives, as the German fighters ruled the air, and the losses suffered by these units were very severe.

On May 12th 1940 at 09:30, it was a Potez 63-11 of GC II/2 which observed and signalled the crossing of the Samois river at Bouillon by enemy motorized columns heading for Sedan. Unfortunately the headquarters, though duly informed, did not react fast enough as strategists were convinced that this only a diversion for an offensive through Belgium.

Heavy losses in the reconnaissance squadrons were not exclusively due to fighters - constant bombing raids against airfields also destroyed a significant number of aircraft. Furthermore, as the operational strips were based not too far from the front lines, they had to be moved at the same time as the German forces advanced. These frequent moves also generated a number of accidents and more attrition. Aircraft that, under normal conditions, would have been rapidly repaired and put back into service were lost during the evacuation of the airfields, given the impossibility of moving them anywhere else. Moreover, this situation made spares virtually impossible to obtain. In this way, over 400 Potez 63-11s were lost during May and June 1940.

For example, GAO 510 carried out 25 missions in six weeks, operated from 14 different airfields and lost six aircraft in operations and six more on the ground destroyed by enemy bombers. Meanwhile GAO 501 received 13 Potez 63-11s in several batches to add to a few Mureaux 115 still in service. Three of the Potez 63-11s were lost in combat (plus 2 Mureaux) while

Opposite and below: *Potez 63-11 planes at the factory airfield in Meaulte captured by the Germans in June 1940.*
Stratus coll.

Unidentified wrecks of Potez 63-11.

Unidentified wrecks of Potez 63-11.
Stratus coll.

Potez 63-11 c/n 570 of GAO 2/520 shot down by German Flak close to Redange.
Stratus coll.

Potez 63-11 c/n 322 at the George V bridge in Orléans. Aircraft of GR I/35 was shot down during a reconnaissance mission on June 10th1940. Fuselage fell from the lorry during transport.

All photos: Stratus coll.

Above and below: *Crash-landed Potez 63-11 c/n 35? of 1. Escadrille, GR I/52.*

eight more had to be abandoned. The human losses amount to 7 crews killed (4 by accident), 3 wounded and 3 prisoners during the May/June combats.

GR II/33 counted successively among its equipment 4 Potez 542s, 13 Potez 637s, 11 Potez 63-11s, 3 Potez 631s and 19 Bloch MB 174s, losing 11 Potez 637s, 4 MB 174s and 5 Potez 63-11s in combat. On the human side, there were 14 dead, 5 wounded and 11 prisoners. This was the price for 186 missions

(including 127 reconnaissance missions in which 12,000 pictures were brought back from the front), totalling 885 flying hours.

In fact, these losses were proportional to the high number of aircraft of this type in service, in units based near the combat zones, and carrying out daily missions. The Potez 63-11 itself proved to be a tough aircraft that could even be dangerous for enemy fighters. For example, on June 9th 1940, a GAO 543 aircraft fought a combat against seven Bf 109s. It was a reinforced armament model (with seven forward-firing machine-guns), and it shot down three of the enemy before being finally shot down after running out of ammunition.

Potez 63-11 c/n 357, GR I/52 destroyed in a crash on May 7th 1941. Aircraft was piloted by Adj/Chef. *Charaud.*

Potez 63-11 on Evreux airfield, occupied by the Germans, June 1940.

BB coll.

Potez 63-11 c/n 373 in German hands.

Luxeuil, a Potez 63-11 of GR II/55 about to take off for a mission.

Lt. *Félix (right) and his crew (Boulard, centre, and Chancrin, left) in front of their Potez 63-11 at Malbouhans in October 1940.*
JG

Potez 63-11 c/n 280.
BB coll.

Below left: Lt. *Gaillard and his Potez 63-11, at Luxeuil in April 1940, barely one month before the French campaign began.*

Below right: Lt. *Félix, CO of 3rd Escadrille in front of a newly delivered 63-11, still with temporary 2-bladed props.*

JG

Potez 63-11 c/n 188 after forced landing.

German troops inspecting captured Potez 63-11
Stratus coll.

Hidden Potez 63-11 at one of the French airfields. Note damaged undercarriage and starboard engine nacelle.
Stratus coll.

Wreck of Potez 63-11 c/n 278. **Stratus coll.**

Sad remains of 63-11 c/n 386 at unknown French airfield, June 1940. **Stratus coll.**

Potez 63-11 c/n 181 after a wheels-up emergency landing.

Potez with an inscription, rather rare on French aircraft in the first stages of the war.

Stratus coll.

After the Armistice

There remained in the French Free Zone some 221 Potez 63-11s, plus 251 others in North Africa and 17 in Syria, while about ten more fled to Great Britain or to Allied territories, to eventually join the FAFL. Two of the latter were in Palestine with the French Fighter Flight *n*°2, based in Haïfa. These were c/n 699 (or possibly 799)/AX673 (ex GAO 1/589, escaped on June 27th 1940) and c/n 670/AX672 (ex GR II/39, escaped July 2nd 1940) which formed this unit together with three MS 406 fighters from Syria. They were detached to Abu-Sueir and took part in the defence of the Nile delta area until spring 1941. Meanwhile, c/n 395/AX 680 and c/n 420/AX691 replaced the former two, which had suffered accidents. C/n 175 and 838 arrived in Great Britain on June 18th and 24th 1940, and were sent to the Middle East in 1942 to join c/n 629 and 817. The French Communication Flight n°3 also flew one of the Potez 63-11s in 1941.

Two Vichy aviation units were in Syria in 1941. In May, while the German and Italian aircraft flying to Iraq were refuelling in the Levant, incidents between the French and British were multiplying. The Potez 63-11s performed reconnaissance flights over British-controlled territories, and two of them were damaged in Damascus during a strafing raid by the RAF. On June 8th, during the ground offensive, GR II/39 and GAO 1/583 respectively had 12 and 5 Potez 63-11s, and three more were delivered during the campaign. Three were shot down, 7 more were destroyed on the ground, and the remaining 10 crashed or were abandoned.

Some 63-11s were eventually flown by the FAFL in the Levant in 1942/43 as part of a scouting squadron, and at the Rayak flying school.

Wing machine-guns are clearly visible on this Potez 63-11.

Pre-flight and souvenir photo before a dawn take-off.

During the engine warm-up, the observer boards his aircraft. One can see the lack of propeller spinners, the wing machine-guns and the Vichy markings on the background aircraft.

The accident of Potez 63-11 c/n 222.

Three Vichy air force Potez 63-11s, fitted with two-bladed propellers for their transfer to the overhaul site, plus one Potez 650. Potez c/n 254 can be seen in the foreground, with open machine-gun fairings showing the ammunition drums.

The Vichy air force in France had a number of 63-11s, which were in service with GR I/14, II/14 and II/22. In North Africa, 63-11s were with GR I/36 in Algeria, and I/52 (reformed) in Morocco, as well as GB II/63 in French Occidental Africa. During the combats against American forces on November 8th-10th 1942, I/52 lost three aircraft shot down and eight more were destroyed on the ground, while I/36 made a few sorties at the cost of three planes crashed. Both North African groups were dissolved in December 1942, and the one in AOF abandoned its aircraft in September 1943.

Reconnaissance Squadron 555 was part of the *Groupe Aérien de Madagascar* at Ivato. In May 1941, its Potez 25s were reinforced with 7 Potez 63-11s, one of which was rapidly destroyed. In May 1942, when the British forces made their landing, the 63-11s performed bombing and reconnaissance missions. Three were shot down, and three more written off on the ground. In November 1942, the Vichy forces had 114 Potez 63-11s, against 236 in November 1941.

Potez 63-11 c/n 670 French Fighter Flight n° 2, Nicosia (Cyprus), January 26th 1941.

FT

A Potez 63-11 of the French Fighter Flight n°2. The under surfaces were split into white and black halves. The rear machine-guns are visible. June 4th 1940.

Pange coll.

Morane MS 406 and Potez 63-11 c/n 420 (AX691) French Fighter Flight in Haïffa at the end of 1941.

FT

Naval Aviation

After the Armistice, the *Aeronavale* unit 4 BR, based in Blida, Algeria, was equipped with Potez 63-11s. The unit had previously operated Martin 167 Marylands, and these were replaced in November 1941 by Potez 63-11s from the *Armée de l'Air*. A total of 9 Potez 63-11s flew with this unit until its disbandment in 1943, and the *Armée de l'Air* recovered those aircraft still in flying condition.

Abandoned Potez 63-11 after crash landing, masked by French staff. **Stratus coll.**

In the *Luftwaffe*

After June 1940, the German forces captured some 80 aircraft still being assembled on the production line at Méaulte. The *Luftwaffe* put them into service for training and liaison purposes. The Méaulte production line was reactivated in 1942 in order to build 120 more aircraft, essentially from the existing stock of parts, and these aircraft were sold in part to Romania, and eventually to Hungary.

In November 1942, the Germans captured the Vichy aircraft, and added some 134 Potez 63-11s to their own fleet.

Potez 63-11 c/n 207captured by the German forces, without engine cowlings or spinners.

Unknown Potez 63-11 in German hands. **Stratus coll.**

Germans on a Potez 63-11.

A German 63-11 ??+QL and Dornier Do 17.

Potez 63-11 modified by the Germans.

A German Potez 63-11 numbered 70.

Right: *SNCAN personnel in front of Potez 63-11 KK+QZ; the aircraft in the background was coded KJ.*

A German 63-11 on a trailer.

Potez 63-11 c/n 501 of GR I/36 landed at Son San Juan (Majorca) on June 21st, 1940 with engine troubles en route to North Africa with Lt. *Doleac,* Lt. *Lecolgnec and* Asp. *Signoret. The aircraft was interned in Spain and was seen here after transfer to Cuatro Vientos (Madrid) at the hands of* Ejército del Aire's Lt. *Gonzalo Hevia. Once in Madrid it was assigned to the flight test unit and stayed in service until 1945. With the French markings removed, it initially sported only the St Andrew's cross on the white rudders, but eventually received full Spanish markings.*

In Romanian service

In April 1942, a first batch of twelve Potez 63-11s was delivered to Romania. These aircraft were still sporting their French identification numbers on arrival, and they were transferred to the 3rd Training Squadron in Brasov, destined for bombing instruction.

Some of this batch were later integrated into the 74th Light Bomber Squadron, which had until then flown Potez 633s. This unit was part of the 3rd Light Bomber Group involved in the Stalingrad battles, where the aircraft carried out attack and reconnaissance missions until spring. The Potez 63s were then withdrawn from the front, in order to re-equip the 3rd Light Bomber Group with Ju 87 D-3 *Stukas* - the unit was subsequently renamed 3rd Dive Bomber Group.

The Potez 63-11s were overhauled before being transferred to the 1st Fighter Wing to be used for night flying training, together with the remaining Potez 633s.

Above: *Sgt. Piturca - Romanian pilot of Potez 63-11.*

Left: *Three Romanian Potez 63-11s, including c/n 101 and 71, with IAR-39 n°12.*

A Romanian Potez 63-11.

Romanian Potez 63-11 c/n 145.

The same Romanian Potez 63-11s and IAR-39 seen on previous page.

In Hungarian Service

by Csaba Becze

Although the Third Reich planned to support Hungary in spring 1943 with 20-25 French aircraft in transport role (Amiot 351/354 or LeO 451), these plans were not realised and Hungary received a significant number of French aircraft only in spring 1944. Twenty Caudron Simouns and 31 Potez 63s arrived to Hungary at this time. The Caudrons were used as liaison aircraft, while the Potez were mostly used as trainers.

The Potez were a mixed batch: fighter, reconnaissance and light bomber aircraft. Their serials are unfortunately unknown. Most of them were in the German KH+?? code series. Theoretically, the Potez 63s received Hungarian markings in the G.701 series, but in fact there is no photographic evidence of a single Potez re-painted in Hungarian colours and markings. The known photos show the Hungarian Potez with German 4 letter codes and German insignia.

The Potez were delivered to Budapest-Ferihegy airfield, to the Hungarian *Repülő Kísérleti Intézet* (Aviation Research Institute, ARI), which evaluated the type first. On April 13th, 1944 the 15th USAAF heavily raided Ferihegy and several of these Potez were damaged or destroyed. Some of the Potez remained in ARI's hands after the evaluation too. The ARI evacuated Ferihegy in late April 1944 after the devastating American air raid on the 13th and they moved to Várpalota airfield. They ferried their Potez here and these aircraft were stored around Várpalota on auxiliary airfields. The ARI was disbanded in November, 1944 and their Potez were not evacuated from the advancing Red Army at the end of the war and most of them suffered serious damage or were destroyed in the heavy ground battles in this area in the winter of 1944/45. Despite these heavy ground battles, a few of them remained in surprisingly good shape.

Apart from the ARI, two other Hungarian flying units used the Potez 63. One of them was the fighter-destroyer combat training unit at Tapolca. They received approximately a dozen Potez, but these aircraft were barely used by them and most of them were withdrawn from this unit by late summer of 1944.

The third Hungarian unit which used the type was the target tug group. For the target towing role, Hungary used mostly the obsolete Junkers Ju 86,

Potez 63-11, ??+HG at Budapest-Ferihegy in the spring of 1944.
C. Becze coll.

Four wireless operators of the ARI on a Potez 63-11 code ??+WU at Budapest-Ferihegy. This particular photo was taken on April 13th 1944 when the US 15th AAF heavily raided that Hungarian base.
Schlosser via Becze

but there were some other second-line types too, including the newly arrived Potez 63. The only target towing squadron was reorganised to group strength in June 1944, with three different squadrons. These target towing squadrons were stationed mostly at Lesvár, Miskolc and Debrecen airfields. In autumn 1944 they retreated to the Csapod area (NW Hungary). On March 27th, 1945 the target tug units got the belated order to retreat to the Third Reich, after the Red Army's spring offensive in the Vienna direction approached Csapod. Because of fuel shortages, the target towing crews destroyed 14 aircraft on the ground, including three Potez 63s, and only a few of their light aircraft managed to reach Zeltweg and Pocking. Perhaps these destroyed Potez were the last Hungarian ones in flyable condition. It is possible that a few target towing Potez were re-painted in Hungarian colours, but there is no evidence for this at all.

Four Potez 63-11s at Ferihegy in the spring of 1944. The first one is KH+TH, the second one is KH+TG.
Schlosser via Becze

KH+TF at Tapolca, summer 1944. The German Balkenkreuz was partially removed from the fuselage, but it was not re-painted in Hungarian colours.
Pocskai via Becze

Detail photos 63-11
Fuselage

General view of Potez 63-11.

Potez 63-11 c/n 318 left by French.

Stratus coll.

Two views of Potez 63-11 c/n 58 with a camera fitted in place of the dorsal machine-gun.

Engine

A Potez 63-11 with two-bladed ferry propellers and the observer's hatch open.

Details of engine of the Potez 63-11 c/n 207captured by the German forces.

The nose of a Potez 63-11.
Stratus coll.

Cockpit

The glazed nose of the Potez 63-11.

Front end of a Potez 63-11.

Nose and canopy of a Potez 63-11.

Open elements canopy of Potez 63-11.

Left: Lt. *Bernard Pelliot sits in a Potez 63-11's cockpit; he'll be killed in action aboard an RAF Halifax later in the war.*

Below left: *Crew climbing aboard a Potez 63-11.*

Below: *Underside view of a Potez 63-11's nose.*

In this picture of the canopy, the door hinges and handle are clearly visible.

Potez 63-11 captured by the Germans at Evreux airfield, June 15th 1940. ***Stratus coll.***

A Potez 63-11 captured by the Germans; note open side panel and observer's hatch. ***Stratus coll.***

Interior of a Potez 63-11's nose seen from the front.

Gunner's position.

Pilot's instrument panel and control column.

Armament

S/Lt. *Bayle about to sit in his cockpit.*

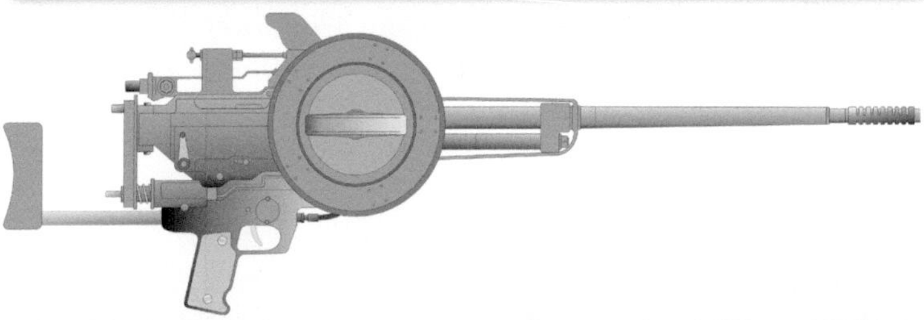

7.5 mm MAC 24 Mk. 1934 machine gun.

S/Lt. *Brune and the MAC 34 machine-gun in the gunner's seat. He was killed with his CO,* Lt. *Rambaud.*

Open underwing fairings of Potez 63-11.
Stratus coll.

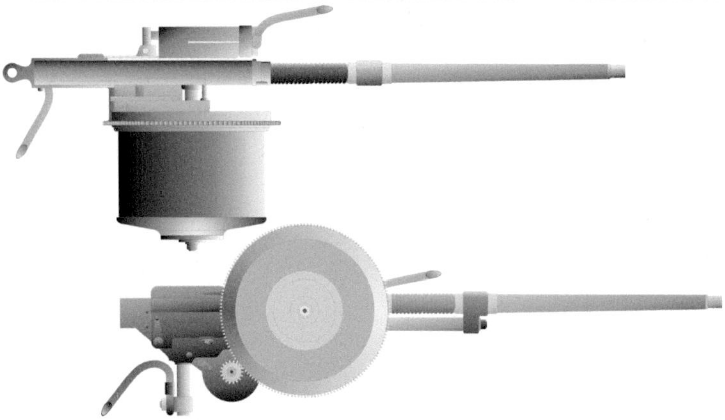

7.5 mm MAC 24 Mk. 1934 machine gun.

Wing machine gun in underwing fairings of a Potez 63-11.
Stratus coll.

Wings and tailplane

The front and aft belly machine-gun fairings, here without armament.

Type and number on the rudder.

The wreck of a Potez 63-11 shows roundel position on wing uppersurface.

A Potez 63-11 with a Bloch 174.
Stratus coll.

A Potez 63-11 with lowered underwing gun fairing; those are hinged at the rear and held in the middle by cables.
Stratus coll.

Specification

Version	Potez 630 C3	Potez 631 C3	Potez 633 B2	Potez 637 A3	Potez 63-11 A3
Wing span	16.00 m 52.493 ft	16.00 m 52.493 ft	16.00 m 52.493 ft	16.00 m 52.493 ft	16.00 m 52.493 ft
Length	11.07 m 38.386 ft	11.07 m 38.386 ft	11.07 m 38.386 ft	11.07 m 38.386 ft	11.07 m 38.386 ft
Height	3.67 m 12.041 ft	3.67 m 12.041 ft	3.67 m 12.041 ft	3.67 m 12.041 ft	3.67 m 12.041 ft
Wing area	32.70 m^2 351.980 sq. ft	32.70m^2 351.980 sq. ft	32.70 m^2 351.980 sq. ft	32.70 m^2 351.980 sq. ft	32.70m^2 351.980 sq. ft
Weight empty	2,450 kg 5,401 lb	2,600 kg 5,732 lb	2,450 kg 5,401 lb	3,154 kg 6,953 lb	3,154 kg 6,953 lb
Weight take off	3,850 kg 8,488 lb	3,760 kg 8,289 lb	4,500 kg 9,921 lb	4,225 kg 9,315 lb	4,433 kg 9,773 lb
Engines	2 radial Hispano-Suiza HS 14Ab 02/03 - 725 mhp/ 715.1 hp, each, from aircraft no egz. HS 14 Ab 10/11 engine	2 radial Gnome-Rhône GR 14 M4/4 "Mars" - 700 mhp/ 690.4 hp, each	2 radial Gnome-Rhône GR 14 M4/4 "Mars" - 700 mhp/ 690.4 hp, each	2 radial Gnome-Rhône GR 14 M6/7 - 710 mhp/ 700.3 hp, each	2 radial Gnome-Rhône GR 14 M4/5 - 710 mhp/ 700.3 hp, each
Fuel	765 l 168 gal	765 l 168 gal	765 l 168 gal	950 l 209 gal	950 l 209 gal
Armament	2 cannons 20 mm Hispano Suiza HS404 1 7.5 mm MAC Mk 1934 machine gun, operated by rear gunner	2 20 mm Hispano Suiza HS9 or HS404 cannons 1 7.5 mm MAC Mk 1934 machine gun, some aircraft also with 4 x MAC Mk 1934 in underwing pods	3 7.5 mm MAC Mk 1934 machine guns 400 kg bombs (8 x 50 kg)	3 7.5 mm MAC Mk 1934 machine guns	3 7.5 mm MAC Mk. 1934 machine guns Bombs: aircraft delivered after 1/1/1940 8 x 10 kg in fuselage +4 x 50 kg under wings
Max speed	440 km/h 273.4 mph	442 km/h 274.6 mph	435 km/h 270.3 mph	435 km/h 270.3 mph	425 km/h 264.1 mph
Rate of climb	9.2 m/s 30.18 ft/s	9.2 m/s 30.18 ft/s	9.2 m/s 30.18 ft/s	9.2 m/s 30.18 ft/s	8.4 m/s 27.56 ft/s
Ceilling	6,000 m 19,685 ft	6,000 m 19,685 ft	8,000 m 26,247 ft	8,000 m 26,247 ft	8,500 m 27,887 ft
Range	1,220 km 758 miles	1,220 km 758 miles	1,320 km 820 miles	1,550 km 963 miles	1,500km 932 miles

Production of Potez 63 family

Type	c/n	Imatricule
Potez 630	3 – 72 73 – 82	C-515 – C-584 X-700 – X-709
Potez 631	1 – 10 11 – 40 41 – 90 91 – 130 131 –155 156 – 207	X-662 – X-671 X-710 – X 739 X-612 – X-661 X-760 – X-799 X-900 – X-924 X-925 – X-976
Potez 633	5C (ex China) 6C (ex China) 7C – 8C (ex China) 21R – 40R (ex Romanian) 12G (ex Greeks) 15G – 24G (ex Greeks)	X-980 X-984 X-982 – X-983 X-740 – X-759 B-232 B-235 – B-244
Potez 637	1 – 60	C-585 – C-644
Potez 63-11	1 – 145 146 – 455	J-804 – J-970 C-646 – C-954

Type	1938	1939	1940	Sub total
Potez 630	44	40	1	85
Potez 631	28	179	2	209
Potez 633	-	36	1	37
Potez 637	-	61	-	61
Potez 63-11	-	257	466	723
			Total	1115

Lyon-Bron base june 1940. Damaged Potez 63-11 c/n 396 (C-895), GAO I/584. In background Potez 631 c/n 161 (x-930) early GC I/1.

BB coll.

Colour profiles

Potez 630 c/n 5 (C-517), 1938.

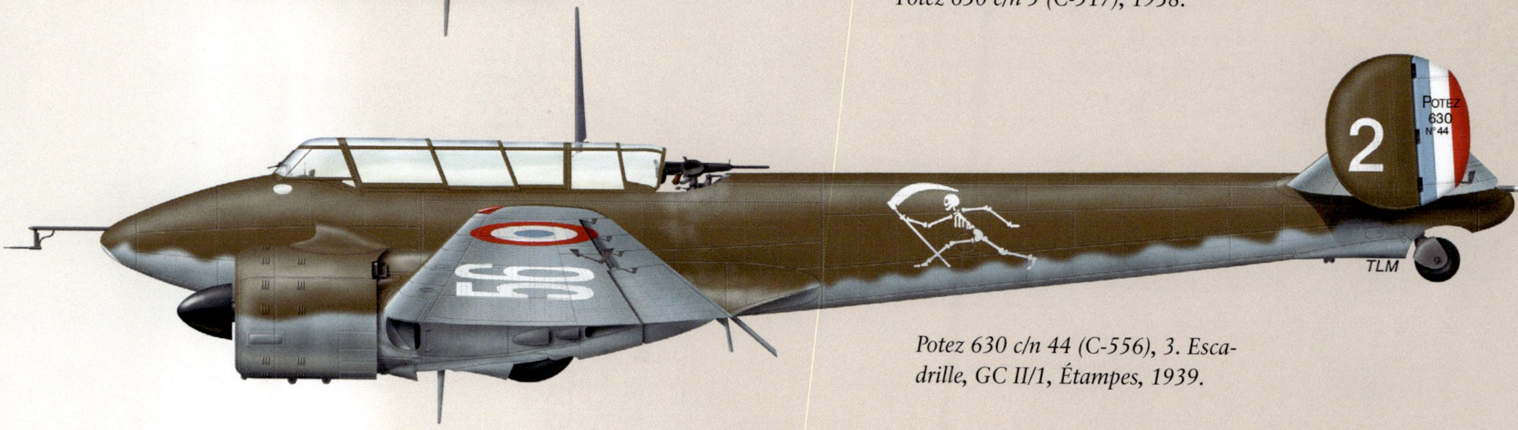

Potez 630 c/n 44 (C-556), 3. Escadrille, GC II/1, Étampes, 1939.

Potez 630 c/n 103 (X-772), 2. Escadrille, GCN I/13, destroyed at Meaux-Esbly (Seine-et-Marne) in an accident September 9th, 1939, two crewmen killed, one wounded.

Potez 630 c/n 75 (X-702), GC I/5, Bordeaux-Mérignac, June 1940.

Potez 630 c/n 75 (X-702), GC I/5,
Bordeaux-Mérignac, June 1940.

Potez 631 c/n 21 (X-720), Escadrille AC1, Aéronautique Navale, Hyères- le Palyvestre, June 1940.

Potez 631 c/n 162 (X-931), ECN 1/13, Moissy Cramayel, end of May 1940.

Potez 631 c/n 49 (X-620), 2. Escadrille, GC I/8, Dijon, April 1940.

Potez 631 c/n 142 (X-911), ECMJ 1/16, Aix-les-Milles, June 23rd 1940.

Potez 631 c/n 169 (X-938), Escadrille AC2, Aéronautique Navale, Calais-Marck, March 1940.

Potez 631 c/n 73 (X-644), ECN 1/13, Meaux-Villenoy, spring 1940.

Potez 631 c/n 73 (X-644), ECN 1/13, spring 1940.

Potez 631 c/n 204 (X-973) Patrol DAT (Défense Aérienne du Territoire), Cne. *des Roziers, Istres, June 1940.*

Potez 631 c/n 205 (X-974), ECN 5/13, *Loyettes, June 1940.*

Potez 631 c/n 58 (X-629), Istres, April 1940.

Potez 631 c/n 47 (X-618), Escadrille AC1, Aéronautique Navale, Cherbourg, June 1940.

Potez 631 c/n 186 (X-955), ECN 4/13, Nîmes-Courbessac, August 1940.

Potez 631 c/n 117 (X-786) ECN 1/13. Unit CO's aircraft in experimental camouflage, January 1940.

Potez 631 c/n 185, GR 3/3, Cognac, 1945.

Potez 631 c/n 198 (X-967), ECN 1/13, flown by Cne. *Treillard.*

Potez 631 c/n 198 (X-967), ECN 1/13 with new night camouflage and Vichy stripes, Nîmes-Courbessac, 1941.

Potez 633 c/n G17 (B-237) ex-Greek.

Potez 633 c/n 21 (X-740),
2. Escadrille, GBA I/51, Étampes,
May 20th, 1940.

Potez 637, c/n 38, 2. Escadrille GR I/52, Aizainvillers, April 20th 1940.

Potez 633 R29 ex-Romanian.

Potez 633, Bach Maï (Tonkin), Col. Pélisse, Groupe Aerien Mixte (GAM).

Potez 637 c/n 3 (C-587), 3. Escadrille, GR II/33, Orconte (Marne), October 1939.

Potez 637 c/n 7 (C-591), 3. Escadrille, GR II/33, October 1939.

Potez 637 c/n 45, (N-629), 1. Escadrille GR I/52, Lt. *Layetz (pilot), Auzainvillers, March 16th 1940.*

Potez 63-11 c/n 357, 2. Escadrille,
GR I/52, 1940.

Potez 63-11 c/n 814 GR I/14,
June 1940.

Potez 63-11 c/n 138 (J-963), DIAP (Depot d'Instruction de l'Aviation Polonaise) Lyon-Bron, March 1940.

Potez 63-11 c/n 138 (J-963), DIAP (Depot d'Instruction de l'Aviation Polonaise) Lyon-Bron, March 1940.

Potez 63-11 c/n 132 (J-657) of DIAP
Lyon-Bron, March 1940.

Potez 63-11, c/n 156, GR II/33,
Athies-sous-Laon, May 10th 1940.

Potez 63-11 c/n 814 (C-655),
1. Escadrille GR I/14, St. Simon - clasters (Aisne), April 1940.

Potez 63-11 c/n 394, GR II/33, July 1940.

Potez 63-11, c/n 501, GR I/36, June 1940.
(See page 172 for the same aircraft in Spanish markings).

Potez 63-11 c/n 288, GAO 511, 1940.

Potez 63-11 c/n 617 GR II/22.

Potez 63-11 c/n 684, 4. Escadrille GR II/22, Vassincourt, May 14th, 1940.

Potez 63-11 c/n 390, 1940.

Potez 63-11 c/n 641, GAO 515, Conantree, June 1940. Lt. Laneyrie, Sgt. Vaillat, S/Lt. Lahaye.

Potez 63-11, GAO 510, French North Africa, 1940.

Potez 63-11 c/n 388 (C-877), GAO 504, Mignault. (Belgium), May 15th 1940.

Potez 63-11 c/n 198, de Buc, 1940.

Potez 63-11 c/n 384 (C-883) GAO 3/551, St-Margueritte-en-Ouche (Eure), June 13th 1940.

Potez 63-11 c/n 217.

Potez 63-11 c/n 287, GR II/14, end of 1940.

Potez 63-11 c/n 366, markings of GR II/14.

Potez 63-11, GAO I/583, Aleppo, Syria, 1941.

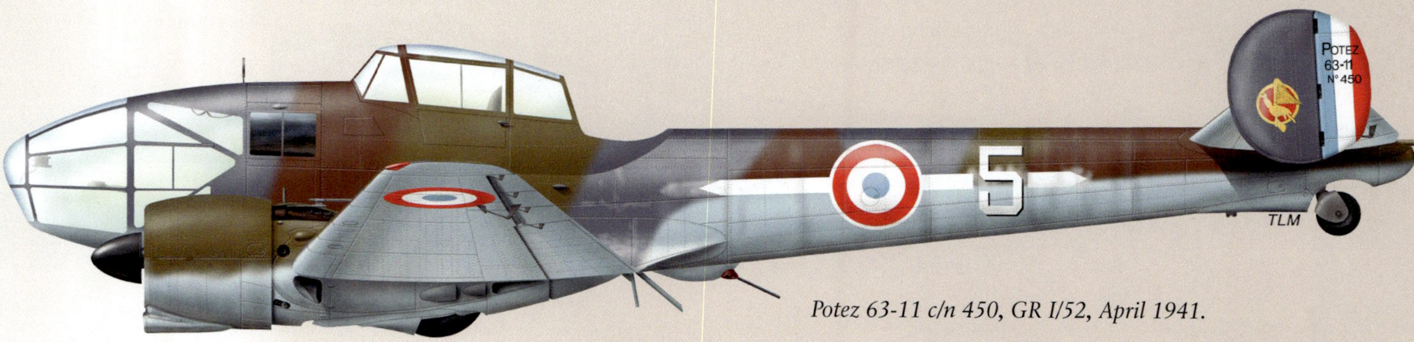

Potez 63-11 c/n 450, GR I/52, April 1941.

Potez 63-11 c/n 232 with a variation of the Vichy markings.

Potez 63-11 c/n 703, GR I/36, end of 1942.

Potez 63-11 c/n 232 (C-731) with a variation in Vichy markings. Groupe Aerien Mixte (GAM) Diégo-Arachaer (Madagascar), May 5th 1940.

Potez 63-11 c/n 12, 2. Escadrille, GR II/22, Montpellier, 1942.

Potez 63-11 c/n 372, GR I/52, Safi (Maroco), November 1942. Cne *Grimal (pilot).*

Potez 63-11 c/n 372 (C-871), Groupe Aerien Mixte (GAM), Diégo-Arachaer (Madagascar), May 5th 1942.

Yugoslav Potez 630 c/n 4 with French ferry registration F-AREY, 1939.

Potez 630 B-1, Troupe d'Aviation Suisse (Swiss Air Force), 1939.

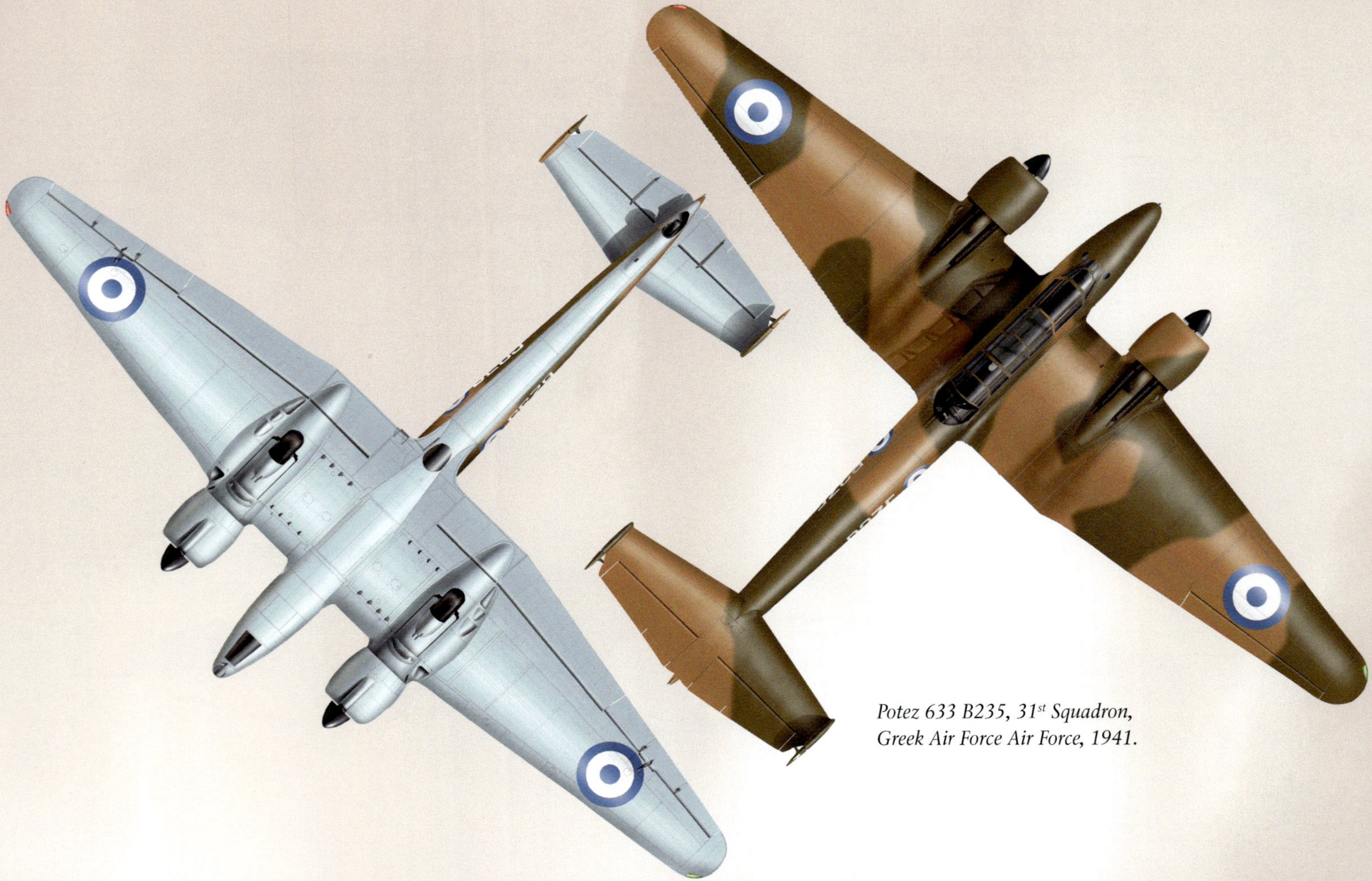

Potez 633 B235, 31st Squadron,
Greek Air Force Air Force, 1941.

Potez 633 B235, 31st Squadron,
Greek Air Force, 1941.

Potez 633 (11), Escadrila 1 Recunoastere,
Romanian Air Force, 1942.

Potez 633 c/n 7 (12), Grupul 3 Bombardament, Romanian Air Force, 1942.

Potez 633 (11), Escadrila 1 Recunoastere, Romanian Air Force, 1942.

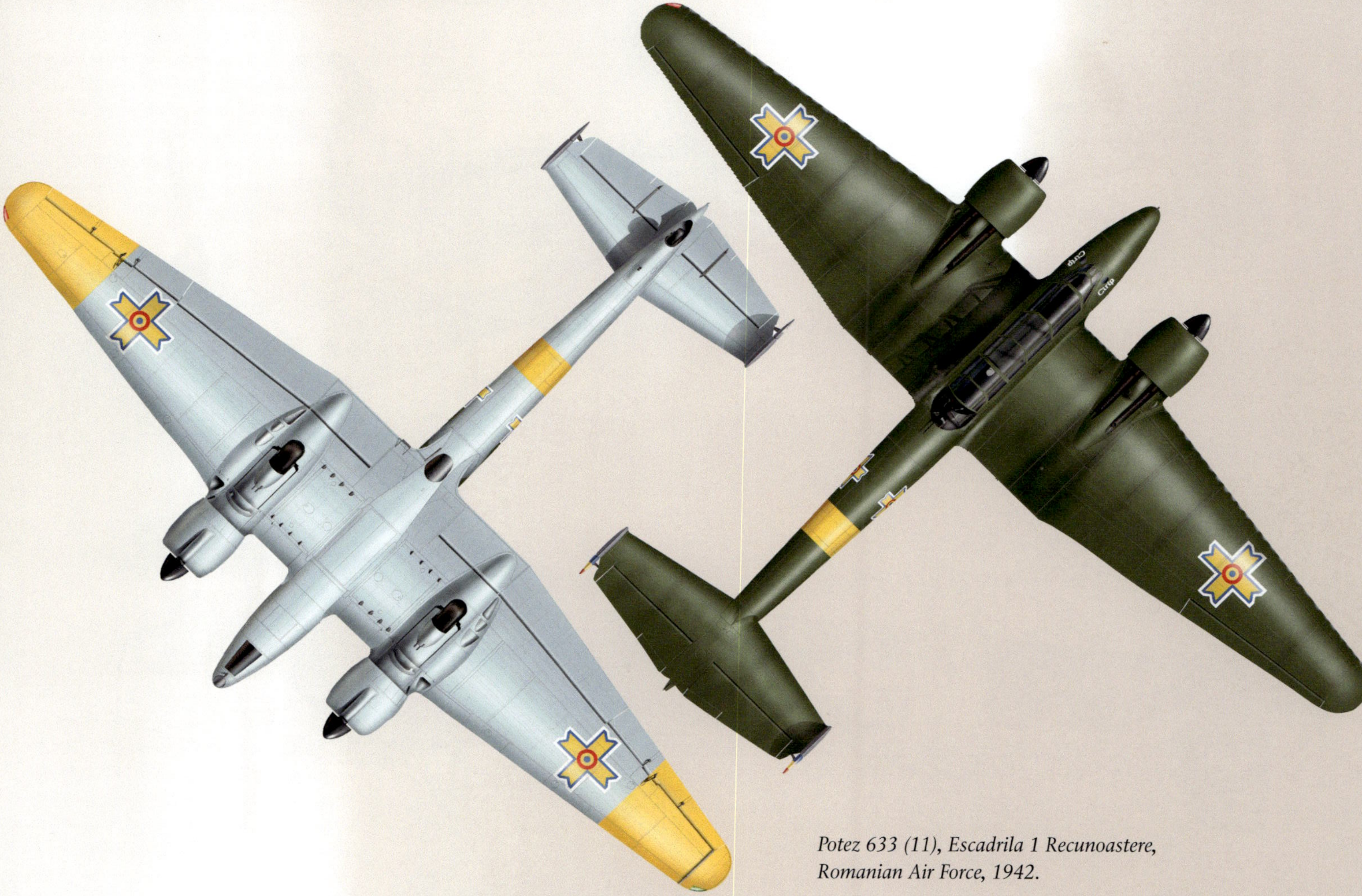

Potez 633 (11), Escadrila 1 Recunoastere,
Romanian Air Force, 1942.

Potez 63-11 (101), Romanian Air Force.

Potez 63-11 c/n 145, Romanian Air Force.

Potez 63-11 c/n 145, Romanian Air Force.

Potez 630 of the Luftwaffe.

Potez 63-11 (70) of the Luftwaffe.

German Potez 63-11 RC+ZF.

German Potez 63-11 (46).

Hungarian Potez 63-11 KH+TF, still in German markings. Hungarian Repülő Kísérleti Intézet *(Aviation Research Institute, ARI, Ferihegy airfield, spring 1944.*

Hungarian Potez 63-11 KH+TH, still in German markings. Hungarian Repülő Kísérleti Intézet *(Aviation Research Institute, ARI, Topolca airfield, summer 1944.*

Potez 63-11 "white 6", Regia Aeronautica,
Piemont, Italy, 1943.

Potez 63-11, Regia Aeronautica,
Mondovi, Italy, 1943.

Potez 63-11, c/n 501, GR I/36, after impression into the Spanish Air Force, Son San Juan. July 1940.

Potez 63-11 c/n 395 (AX 680), French Fighter Flight n° 2, Haïffa (Palestine), October 1940.

Potez 63-11 c/n 838, GR I/4 Oldham UK, 1940. Lt. Jacob, S/Lt. Nuemann, Sgt/Chef. Morell.

Potez 63-11 c/n 175, Pengam Moores near Cardiff, UK, July 1941.

Potez 63-11 c/n 670, French Fighter Flight n° 2, Ismailia, Egypt, July 1940.

Potez 63-11 c/n 670 (AX673), French Fighter Flight n° 2, Ismailia, Egypt, December 1940.

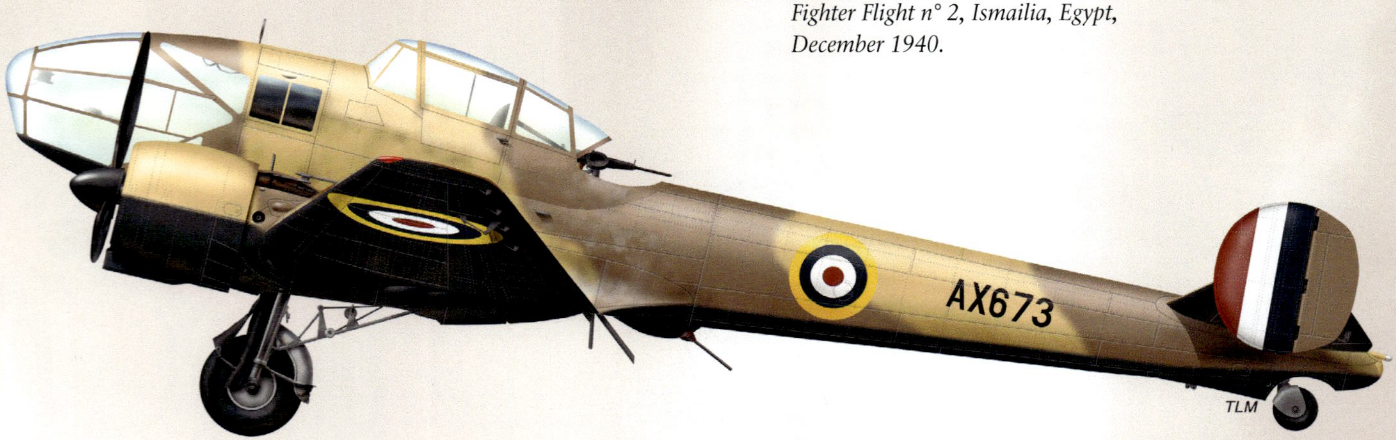